MW00616503

HOW BIG
IS BIG AND
HOW SMALL
IS SMALL

The Sizes of Everything and Why

by

Timothy Paul Smith

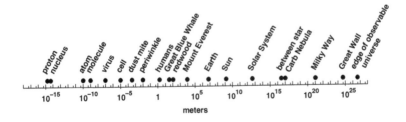

OXFORD
UNIVERSITY PRESS

OXFORD

UNIVERSITY PRESS

Great Clarendon Street, Oxford, OX2 6DP,
United Kingdom

Oxford University Press is a department of the University of Oxford.
It furthers the University's objective of excellence in research, scholarship,
and education by publishing worldwide. Oxford is a registered trade mark of
Oxford University Press in the UK and in certain other countries

First Edition published in 2013

Impression: 1

Published in the United States of America by Oxford University Press
198 Madison Avenue, New York, NY 10016, United States of America

British Library Cataloguing in Publication Data
Data available

Library of Congress Control Number: 2013937705

ISBN 978–0–19–968119–8

Printed and bound by
CPI Group (UK) Ltd, Croydon, CR0 4YY

Contents

List of Figures   iv

List of Tables   vii

1. From Quarks to the Cosmos: An Introduction   1
2. Scales of the Living World   14
3. Big Numbers; Avogadro's Number   30
4. Scales of Nature   45
5. Little Numbers; Boltzmann's and Planck's Constants   63
6. The Sand Reckoner   78
7. Energy   94
8. Fleeting Moments of Time   110
9. Deep and Epic Time   127
10. Down to Atoms   146
11. How Small Is Small   164
12. Stepping Into Space: the Scales of the Solar System   184
13. From the Stars to the Edge of the Universe   205
14. A Little Chapter about Truly Big Numbers   228
15. Forces That Sculpture Nature and Shape Destiny   239

Index   249

List of Figures

1.1 The original definition and measurement of the meter. 6

1.2 Towns embedded in counties, embedded in states, embedded in nations. 9

2.1 The relationship between surface and volume. 19

2.2 As towns grow into cities the traffic for supplies must increase. 20

2.3 A collection of *Littorina saxatilis* (periwinkle) collected in Scotland. 23

2.4 Savanna elephant *Loxodonta african*, *Loxodonta catherdralus* and Notre Dame de Paris. 25

3.1 Loschmidt's method for determining the size of a molecule. 40

4.1 The Beaufort and Saffir–Simpson scales and windspeed. 48

4.2 The Richter scale versus the Mercalli scale. 51

4.3 The number of hurricanes in the North Atlantic (2003–2012) as a function of windspeed. 51

4.4 The Mohs scale versus the absolute hardness of minerals. 53

4.5 The stars of Orion as reported in the *Almagest*. 55

4.6 The frets and related frequencies of a guitar. 62

5.1 Reversible and nonreversible collisions on a pool table. 69

5.2 Order as shown with a die and pennies. 70

5.3 The distribution of light described by Planck's law. 74

5.4 The Planck length. 77

6.1 Eratosthenes' method for measuring the size of the Earth. 84

6.2 Aristarchus's method for measuring the size of the Moon. 86

6.3 Surveying the distance to unreachable points with angles and baselines. 88

6.4 Aristarchus's method for measuring the size of and distance to the Sun. 88

6.5 The size of a geocentric and a heliocentric universe. 90

7.1 Energy densities for chemical and nuclear fuels. 101

7.2 The energy that is released when hydrogen burns, at the molecular level. 105

7.3 The octane molecule. 105

7.4 Energy from nuclear fission. 106

8.1 The *analemma*, as viewed from New York City. 115

8.2 The discovery of the omega particle. 124

9.1 Solar versus sidereal time. 132

9.2 The decay of carbon-14 used for dating. 137

9.3 The layers or *strata* of the Grand Canyon. 139

9.4 The supercontinent *Pangaea* about 200–300 million years ago. 142

10.1 The cross section of a standard meter bar from the 1880s. 149

10.2 A marching band performs a wheel and demonstrates interference. 152

10.3 Laser light on a single hair shows an interference pattern. 152

10.4 Trees in an orchard demonstrate the facets in a crystal. 157

10.5 The interference pattern of X-rays from a crystal and DNA. 158

10.6 Wavefunctions of the hydrogen atom. 161

10.7 The periodic table of the elements. 162

11.1 Detail from a small part of the table of nuclides. 166

11.2 The table of nuclides. 167

11.3 The nuclear force. 172

11.4 An image of a nucleus made of protons and neutrons with quarks inside of them. 173

11.5 A collection of different types of particles made of quarks. 176

11.6 The interaction between a proton and neutron in terms of pions and quarks. 177

11.7 The interaction between quarks in terms of gluons and color charge. 178

11.8 The electric charge distribution of a neutron. 179

12.1 Kepler's third law demonstrated by Jupiter's moons and the planets. 191

12.2 The transit of Venus as seen by two separate observers. 192

12.3 Some of the moons in our solar system. 195

12.4 The rings of Saturn. 198

12.5 The Titus–Bode law. 200

13.1 Parallax of a near star. 209

13.2 The H–R diagram of star color versus luminosity.   213
13.3 A map of the Milky Way.   217
13.4 The Andromeda galaxy.   219
13.5 The Doppler effect in water waves.   221
13.6 Comparison of the size of galaxies, group, superclusters and the universe.   223
13.7 The large-scale structure of the universe.   225
14.1 Everything is small compared to infinity.   235
15.1 Different forces of nature dominate different scales.   245
15.2 Proton–proton fusion.   246

List of Tables

3.1 Names of numbers. 31

6.1 The symbols and values of numerals in the Roman number system. 81

6.2 The symbols, values and names of numerals in the Greek number system. 82

6.3 A comparison of the astronomical measurements of Aristarchus, of Archimedes and of modern astronomical measurements. 92

7.1 Comparison of nuclear and chemical forces and bonds. 107

7.2 Comparison of the four fundamental forces in nature. 107

8.1 Exotic subnuclear particles; lifetimes, forces and decay. 123

9.1 Comparison of decimal and standard time. 128

9.2 Days and years of the planets. 133

9.3 Present geologic time. 140

9.4 Supercontinents through the history of the Earth. 143

12.1 Radius of the orbits of the planets: Copernicus vs modern measurements. 189

12.2 Some properties of the planets. 199

12.3 Some properties of the dwarf planets. 203

13.1 List of nearby stars. 211

14.1 Examples of sets that have an infinite number of members. 237

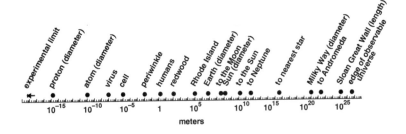

❧ 1 ❧

From Quarks to the Cosmos: An Introduction

If the meter is the measure of humans, then we are closer to quarks than we are to quasars. However, if we take the second as the heartbeat of our lives, then we are closer to the age of the universe than to the lifetime of elementary particles. There are well over forty-five orders of magnitude between the largest things we have ever measured—the grand breadth of the universe itself—and our smallest measurement, the probing of those iotas of matter, quarks, electrons and gluons. There are also over forty orders of magnitude between the fastest events clocked and the slowest events, which we continue to watch evolve and unfold. Somewhere in the middle of these ranges are the scales where we humans spend our lives, the scales measured in meters and seconds. The breadth of these scales is truly astonishing and it is a credit to modern science and to our intellectual capacity that we can write down forty-digit numbers with certainty and a fair amount of precision.

This book is a guide to understanding and appreciating those numbers, both the large distances and the small. A concert or an art exhibit may be beautiful and awe-inspiring, and quarks and the cosmos are awe-inspiring and, I would argue, beautiful too. But the concert or exhibit may be better appreciated with program notes or a guide. Likewise, with diligence and a good guide, we can learn what these vast and infinitesimal numbers mean. We can internalize and appreciate their beauty.

The universe is about 10^{27} m across. That is a 1 followed by twenty-seven zeros: 1,000,000,000,000,000,000,000,000,000 m. At the other extreme, we have probed neutrons and protons. We know that they contain quarks and we know that the quarks are smaller than 10^{-18} m. That is a 1 preceded by seventeen zeros: 0.000,000,000,000,000,001 m. But these are numbers with which we have no connection in our everyday experiences. These numbers are bigger than the national debt, or even the gross national product of all the nations on Earth expressed in pennies. How are we ever going to understand these monstrous numbers? We will do it with analogies and by developing an appreciation for what scales mean. We will also look at systems and slices of nature the scales of which have a smaller range. We will build up, with small steps, to these massive scales. But you should not be disappointed that we do not start out with the ultimate of scales, for it is a grand tour to get there. It is a journey through atoms, sequoias, the Sloan Great Wall, pianos, whales, quarks and rock concerts. On the way we will encounter geniuses and madmen, surveyors and seismographers, horses and hurricanes.

Humans can measure things that are 10^{-18} m across. If we were to count how many of these tiny things could be placed side by side across the universe we would end up with a number with forty-five digits! But that is only one of the endpoints, one of the brackets that hold all of nature. It is not the starting point. We start, curiously enough, with what we understand and use every day. We start with the measuring stick of man.

***

Humans have developed a number of different standards to measure our world, and in general these standards have reflected our own stature, or our labors or our lives. The second is about the time between heartbeats. The traditional acre is related to a day's labor. The day itself may be astronomical in origin, but its importance lies in how the Earth's rotation affects our lives by telling us when to sleep or have breakfast. The pound is a handful of dirt and the cup is a good size for a drink. The traditional units of bushels, pecks, gills and gallons are all useful units for measuring things we use every day.

However, it is the length scale that is of primary interest when measuring the universe, and we humans have been prolific when devising standards by which we can measure the length of things. The Romans measured long distances with the mile, which literally means "one thousand paces" (*mille passus*), where the pace is two steps or the distance from the right footstep to the right footstep. This is a practical unit for marching off distances with an army.

More ancient than the mile is the *stade*. The stade has an important distinction in athletics: it is the distance of the ancient Olympic footrace. We know the Olympic stade is 192.8 m, since we can still go to Olympia and measure where the races were held. It is from the stade that we derive the name of the place where races, and now other athletic events, are held: the stadium. In the history of astronomy, the stade also has an important role, for Eratosthenes (276–194 BC), the Greek mathematician and astronomer calculated the size of the Earth and obtained a circumference of 250,000 stades. Unfortunately for us, it is not clear which stade he used. Was it the Olympic stade, the Roman stade of 185 m or another unit: the *itinerary stade* of 157 m? In any case, the distance he calculated of 39,000–46,000 km is amazingly close to the 40,100 km modern measurements yield. Still, the mile or the stade may be good standards when measuring the length of an army's march or the breadth of an empire, or even the girth of our home planet, but they are not the units we use when we measure ourselves. When we measure ourselves and our homes we use the foot, the yard or the *cubit*. Every single one of these units starts out with some part of the human body, as well as a story which comes out of the misty past.

The cubit is one of those ancient measurements that is very convenient for humans. It is the distance from the elbow to the outstretched fingertips. We all come with a built-in cubit measuring stick. It corresponds to about 18 in or 45 cm. We are told that Noah's Ark was 300 cubits long, 50 cubits wide and 30 cubits high, and the Ark of the Covenant was one and a half cubits high and wide and two and a half cubits long. But the ancient world did not have the ISO (International Organization for Standardization) watching over it and, much like the stade, there was a multitude of cubits. An important and well-studied cubit is referred to as the "Egyptian Old Royal Cubit." In this particular case we know from ancient documents that the Great Pyramid of Giza is 280 royal cubits in length and, since the pyramid is still there, we can go out and actually measure it, and so we know that this royal cubit is equal to 52 cm.

Here we can see the start of a problem. The cubit may be a very convenient unit and any master shipwright or stone cutter can offer his forearm as a standard. But if the stone blocks for the pyramid are to be, let us say, ten cubits on a side, and one quarry uses 45 cm and another quarry uses 52 cm, they will not fit together at the building site. So civilization developed standards. We may like the story of King Henry I of England offering his foot as the royal standard foot (he really was not tall enough to have a foot about 30 cm long), but far more useful was the iron bar mounted in the wall of the marketplace with standard lengths (at least for that market) inscribed on them.

And then we move into the age of the Enlightenment and the metric system. The idea was that this new standard, the meter, would not depend upon a particular man, for after the French Revolution there was no king left who could offer his royal foot as a standard. For all men are created equal, even if we all have our own unique stature. So the meter was to be based on something outside of a simple human dimension, something universal. It should also be based on something that all humankind had access to, such that anyone could go out and create their own meter stick. Thus 10,000,000 m was defined as the distance between the equator and the poles of the Earth. What could be more noble and permanent, more universally accessible than the planet upon which we stand?

Establishing an ideal meter is all well and good, but in a practical sense, how does one go about creating a meter stick? How far is it from the equator to the pole and how did we measure this distance in 1792, or at any other time? No one would visit the North Pole until Robert Edwin Peary's team did so in 1909, and the continent upon which the South Pole is located was still *terra incognita*. In fact Antarctica (*Terra Australis*) was not even sighted until the crew of the Russian ship *Vostok* saw it in 1820. The first expedition to reach the South Pole, led by the Norwegian Roald Amundsen, did not reach it until 1911, and they were not measuring the distance there with meticulous detail.

So how do we establish the meter? What we would like to do is pace off the distance from the equator to a pole. Imagine that I start in Ecuador and march north across the Andes mountains, then through the jungles and deserts of Central America and Mexico, always keeping Polaris, the north star, in front of me. On across the plains of the United States, the boreal forest of Canada and finally the tundra and ice of the Arctic to the North Pole. After my epic trek I find (let us suppose) that

I have taken 12,345,678 steps (left foot to left foot is two steps). Then 12,345,678 steps is 10,000,000 m, and after a bit of division I find that 1.2345678 of my steps is 1 m. I can now mass-produce my meter stick!

The beauty of this system is that if we are careful, your stride does not have to be the same length as my stride. In fact you could replace the steps with revolutions of a bicycle wheel or a surveyor's chain and you and I would end up with meter sticks of the same length. If the Earth was smooth and regular you could perform your measurement someplace else, such as through London or Paris or even through Thomas Jefferson's home at Monticello. You could even stride off to the South Pole and you would still end up with the exact same meter.

But there is a problem with this technique, and it is not just the mountains, canyons, swamps and rivers that will mess up our uniform stride length—a good surveyor can measure across all of these. The problem was that no one had gone to either pole until over a century after the meter was established. Yet the French Academy of Science, which was the organization that established the meter, understood this limit. So when they sent out their two survey teams, headed by Jean Baptiste Joseph Delambre and André Méchain Pierre François, they were only to measure the distance between Barcelona and Dunkirk and their respective latitudes (see Figure 1.1). The reason that this works is that if you can only measure a fraction of the equatorial–polar trek, but you know what fraction you have measured, you can calculate the length of the whole trek. For example, if you measure the distance from 36° north to 45° north you have measured 9° out of 90°, one tenth of the whole distance, or 1,000,000 m.

Now nothing is ever as simple as it seems when proposed, and the survey across France took seven years to complete (1792–1799). The measurement was hard, the Earth is far from being uniform and the French Revolution was going on. But ideally anyone can establish a meter stick. All you need is to measure a quarter circumference of the Earth and divide by 10,000,000.

***

Why divide by 10,000,000? Why not divide the equator–pole distance by a simple number like 1? Or why not use a more common metric number like a thousand or a million? The answer is that by using ten million we end up with a meter that is of a useful human scale. The meter is

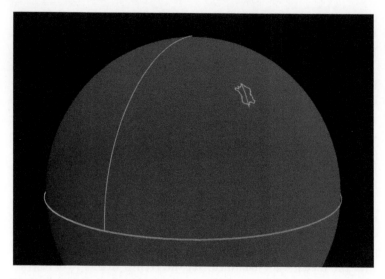

Figure 1.1 The relationship between surface and volume. From the equator to the pole is 10,000,000 m. Jean Baptiste Joseph Delambre and André Méchain Pierre François measured the distance across France, between Barcelona and Dunkirk, to establish the original meter.

about 2 cubits, and it is just over one yard long. It is a very human-sized length. It is interesting to look at another definition of a meter that the French Academy of Science considered. They toyed with the idea of defining a meter as the length of a pendulum with a period of oscillation of 2 sec. However, it turns out that the gravitational attraction of the Earth changes slightly at different latitudes. That would mean that a "meter" established near the equator would be different from one established near the poles, which is not a good way of defining a standard. But in either the Earth-size meter or the pendulum-based meter there were always the curious factors, the 10,000,000 and the 2 sec. Why? The academy already knew approximately the size of the Earth and the length the pendulum would be, and both of them gave a useful, "human-size" meter. A pendulum with a period of 2 sec is 99.4 cm long. The pendulum definition would have led to a very similar meter to the one we have today.

Throughout this book distance will usually be described in meters and time in seconds. These are the working units of modern science.

But for all practical purposes, when you see "meter" you can think "the size of a human" and when you see "second" you can think "a heart beat."

***

Many disciplines within science have their own specialized units that we will bump into. In atomic physics a measuring stick the size of an atom would be useful. So we have the length unit called the *angstrom* (symbol Å). The angstrom is roughly the size of a hydrogen atom. It is exactly 10^{-10} m, or one ten billionth of a meter. Why ten billion and not a billion? Because it is very nearly the size of an atom and, as we will see later, it is very useful. In nuclear physics we use the *fermi* (symbol fm), which is about the size of a proton or a neutron. It is defined not by the proton or the neutron, but by the meter. A fermi is a quadrillionth of a meter (10^{-15} m). It is incredibly small, but it is still defined in terms of the human-scale meter.

Astronomy uses three non-metric units: an astronomical unit (AU): a light-year (ly) and a parsec (pc). An astronomical unit is the distance between the Sun and the Earth. A parsec is defined by the technique for surveying the stars, which we will consider in Chapter 13. A parsec is also roughly the distance between neighboring stars. Finally, the light year reminds us that even if light is very fast, space is that much bigger. But even in astronomy we can, and in this book, will, write stellar, galactic and even cosmic distances in terms of meters.

***

One of the keys to the success of science has been the way scientists organize their observations and data into systems. Carolus Linnaeus (1707–1778), the Swedish naturalist, said that the first step to understanding nature is to uniquely and permanently name the species. But what may be more important is that he then went on and organized animals and plants into his famous binomial system, where a species belongs to a genus, and a genus to a family, an order and so forth. No longer was the world of the naturalist just a jumble of species, but rather there is an organization, a branching system that tells us about connections and relationships between species.

In this book's presentation of the size of things we will also need a system, for if we just pour out interesting but randomly organized facts it will be a confusing jumble. If I sandwich the description of the life-time of an exotic particle called the muon between the size of Jupiter and the energy required for an ant to do a push-up, this would not help us understand how big is big or why things are of a certain size. So in this book we will generally work from scales we know, such as the me-ter or the second, to the scales we are not used to, such as the size of galaxies or the lifetime of an exotic subatomic particle. This does have the problem of making it seem like things are arranged physically next to each other, which is not really right. Phenomena of different scales can be embedded in each other; atoms in molecules, molecules in cells, planets in galaxies. And this is useful to us because this embedding helps us recognize when we have taken a true step in nature. Let us consider one more example. Even if the Earth is much larger than Mercury, it is on the same organizational level because Mercury cannot be contained within the Earth. These planets are embedded within the solar system, and so are on a different level from the solar system. This technique of looking to see what objects can be embedded in other objects will not always help us organize things, but in many cases it can guide us. This may sound a bit too abstract, so let me draw an example from the world of humans.

Humans created types of governments that are embedded in each other and act differently at different levels because of the way they solve the problem of how to govern. I live in a small town in New England, nestled between the Green and the White Mountains. The town's job is to maintain roads, especially those that only connect different parts of the town. The town also has school, police, fire and recreation de-partments. Once a year we gather at a town meeting and debate issues, vote on budgets and impose taxes on ourselves to pay for what we think the town should do. We have commercial centers and a civic life with churches, clubs, schools, sports teams and places to gather to work or learn. These social facets may in fact be a better definition of the town than the government, but that there is a close parallel is not surprising since New England towns grew out of parishes and markets and not the other way around. At the time our town was settled, most people could walk to the town center in about an hour, which is a reasonable amount of time to invest in getting to the market or the church. This limit of an hour's walk means that the size of the town is about 10 km or half a dozen miles on a side. This size in fact became the standard across much of the United States and Canada in the form of the *township and range*, or

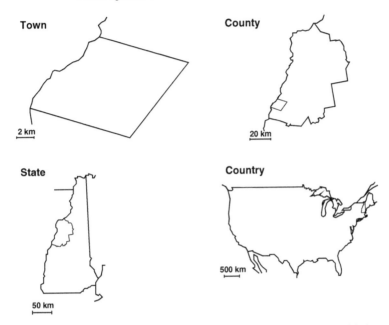

Figure 1.2 Towns embedded in counties, embedded in states, embedded in nations.

Public Land Survey System, which was endorsed by Thomas Jefferson. This system fitted the size of the town to the needs of its citizens.

The next size of government is that of the county (see Figure 1.2). Whereas counties have many functions, including sheriff's departments, roads, parks and social support agencies, it is most likely the county courthouse that you might need to visit; not to see the judge, but rather the county clerk. For this is the place where land transactions and other important events are registered and the records maintained. In Jefferson's public land survey system the county is six townships from east to west and six townships from north to south, or 36 miles — 60 km — on a side. This means that it is not more than about 20 miles or 30 km for the average citizen to travel to the clerk and courthouse. To travel there on foot or even horse is an all-day event. The place is accessible, but not on a weekly basis.

In many places the role of the town, the county, or the city are different from what I am describing here, but the point of this exercise is that different systems can embed themselves in each other. Also, we note that the larger the area that the government unit covers, the more remote and abstract is the interaction with the individual.

When we move to the next level of government—the state—it plays a very different role than towns or counties, and I also interact with it in a very different way. To start with when we ask the question, "What is the natural size of a state?" we see a great range of solutions, from the small New England states like Rhode Island or New Hampshire, to the mega-states like Texas or Alaska. Their boundaries are often defined by geographic constraints, such as rivers and oceans, or by history. Curiously enough, we will see that galaxies are similar in this respect. Moreover, the mechanisms of interaction are new. I have no reason to go to the state capital. I can send things by mail or email, or if I need to go to the Department of Motor Vehicles, I can go to a satellite office. My input is through my state representative. This means that the size of the state is not simply limited by how far I can travel. The state still deals with serving individuals. The state licenses you to drive or be married; it licenses certain types of business; it charters certain types of organizations and sets standards for schools and restaurants. It also legislates the rules by which towns and counties operate. But it is dealing a bit more with other levels of government, and a bit less with individuals.

The federal government is even more remote from our lives than the state. It may dominate the news, but it really is less of a factor in our everyday lives. I carry my state driver's license in my wallet every day, but I only carry my passport on the rarer days that I travel internationally. My method of input to the federal government is even more remote. I am represented by people I have never met and who sit in a legislature a thousand kilometers away. An educational policy set by the federal government affects a few days of school a year, but the state sets minimum standards for all classes, and it is the budget of the town and the policies of the local board that determine when schools start every day and who is teaching what. The federal government deals most directly with other governments at its own or similar levels, with states, other nations and the whole planet.

Finally there is the United Nations, which is an organization and not truly a government, for its members have not relinquished their sovereignty. It is a meeting place of nations, a forum where the world can meet together and work out solutions to common problems.

This nested system of governments I have just described—town, county, state, nation and international organization—is not the only way that governments could be organized. Most readers will not live in a place like my town. They will point out that I have missed cities, villages and boroughs, or that the state could be replaced with a province, a district, or a territory. Also, the balance between the responsibilities of different levels of government is continuously in flux. For example, with the recent establishment of a Welsh Parliament in Cardiff and a Scottish one in Edinburgh and the increasing importance of the European Union, will the role of the United Kingdom change? (Will England get its own Parliament?)

Government levels may be in flux, and particular types of government are not universal, but there are certain general trends. Units of government have a more direct effect on units of government of a similar magnitude. Also, it is very unusual to get a government that is embedded in more than one other government. Although there *are* school districts that cross state lines this is not common or simple. Nature tends to divide itself with more firmness and distinction. Atomic, planetary and galactic systems are well separated in magnitude. Systems that neighbor each other in size do affect each other. If I want to understand chemistry it would help to understand atomic physics, but it would not help as much to study nuclear physics or cosmology. So in this book, as in governments, we will see the embedding of systems, and the effect that different levels of nature have on each other.

***

At the beginning of this chapter we introduced the meter, and argued that it was a unit we could understand and that we could do this because it is the size of a human, but so far we have not really measured anything. We talked about systems that can be embedded in bigger systems, using as an example the different levels of government, but we have not really talked about the different levels of nature. Describing all the levels of nature is a huge task, which is beyond the scope of this chapter, but it would be a disappointment to stop here without at least sketching out the size of the universe and quarks. Envisioning forty-five orders of magnitude is a great mental challenge, and we will spend the rest of this book learning how to deal with these numbers, but I think we deserve a quick tour of nature right now.

We start with the meter, the measuring stick of our everyday lives. The average adult human is about 1.7 m tall, and we tend to pass the 1-m mark in height when we are in kindergarten. Chairs are half a meter up and beds are about 2 m long. Meters measure things that fit us.

On this quick tour we will now jump to things that are one hundred thousand times bigger than us. What things are 100,000 m (10^5 m) or 100 km in length? This is the distance you can drive in an hour. It is also the distance across a state like Connecticut or a country like Wales or the Lebanon.

At 100,000 times the distance across Connecticut (10^{10} m), we are looking at the world at the planetary level. This is 26 times the distance from the Earth to the Moon, but only 7% of the way to the Sun. All the problems of state versus nation seem insignificant at this scale.

Another jump of 100,000 times (10^{15} m) brings us to a scale bigger than our solar system, a chunk of space encompassing all the planets and even part of the Oort cloud, the belt of comets that surround our star system. It takes light 40 days to travel this distance.

Now a jump of another 100,000 times (10^{20} m) brings us to galactic scales. 10^{20} m is 3.3 kpc (kiloparsecs). Our galaxy is about 10^{21} m across.

Another jump of 100,000 means that we are no longer looking at individual galaxies, but rather at distributions of galaxies. At 10^{25} m across is the Sloan Great Wall of galaxies.

Finally, one more jump brings us to 10^{30} m. This is bigger than the whole of the observable universe. The edge of everything—and this really is a difficult idea—is 10^{27} m "out there." Ten followed by 27 zeros. And beyond that?

***

What about the complementary tour? How small is small?

One hundred thousand times smaller than a meter (10^{-5} m) is the world of cells. Cells come in a range of sizes from the ostrich egg, which can be up to 15 cm long, to the bacterium, which is a millionth of a meter in length. But the red blood cell, at 10^{-5} m, is often cited as typical. In the light of our recent tour of big, we note that a red blood cell is as much smaller than our whole body as a person is smaller than Connecticut. So you can think of that single cell wandering around the body as much like a person wandering in Connecticut. Actually the analogy is not quite right, for a human wandering tends to stay on the surface of the Earth, whereas the cell can circulate through the whole three-dimensional body.

One hundred thousand times smaller than the red blood cell (10^{-10} m) is the atomic world. The diameter of the simplest atom, hydrogen, is 1.06×10^{-10} m. Heavier atoms, such as gold or lead, can be 200 times more massive, but only three or four times the diameter.

One hundred thousand times smaller than an atom (10^{-15} m) is the world of nuclear physics. The proton is 1.7×10^{-15} m across. At this distance the electromagnetic force, which holds together atoms and molecules, is unimportant. Gravity, the force which holds together galaxies, is even less important than the electromagnetic force. Both of these are dwarfed by the *Strong Force*, the force which holds together nuclear matter. This is a force we ignore in our everyday lives, even though it is so very powerful and it is embedded in every iota of what we are made of.

One more step of one hundred thousand and we fall off the other end of the scale. We do not know what the world looks like at 10^{-20} m. This limit is a bit different than the 10^{27} m edge of the observable universe. Here we are limited by our biggest experiments. When at the beginning of this chapter I said that the size of a quark is less the 10^{-18} m, I meant that this is the limit of what we can see with our biggest accelerators. We have good theories that lead us to believe that quarks may be much much smaller. They may be as much smaller than us as we are smaller than the universe, but we really do not know. A jump from our present experimental limit to this ultimate *Planck scale* is like jumping from the distances between stars to distances that span the whole universe and ignoring galaxies and galactic structure along the way. There could be a lot of interesting stuff between our smallest present measurements and this ultimate Planck scale; we just do not know.

***

So are we closer to quarks than to quasars? That is a hard question to answer, for the answers seem to be full of all sorts of caveats. Things like "What do we mean by closer?" and "What are these experimental limits?" And, by the way, since we are asking tough questions, "How in the world am I supposed to understand a number with forty-five digits?"

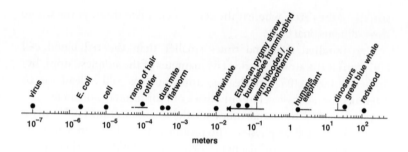

2

Scales of the Living World

The voyage down to quarks and up to stars and galaxies is epic. But more interesting than simply "How big are things?" is the question of why are things the size they are. Why does scale matter? Why are there no Lilliputians or giants? Why it is that we cannot have a solar system inside of an atom? As in any voyage of exploration we need to do a lot of preparation in the next few chapters, but first we will embark upon a short cruise around a familiar neighborhood. We will look at the scales of the living world.

***

The living world still has an impressive span, from things that measure a millionth of a meter to those that are a hundred meters in length. Lifetimes range from 20 min for bacterial division to almost 5000 years for the Great Basin bristlecone pine *Pinus longaeva*. Life covers eight orders of magnitude in both time and space, and someplace in the middle are humans. So this will all feel familiar and we will be very much at home.

The most important tools in science are our eyes and our brain. We have five senses, but the vast majority of the information we have about the size and shape of our world comes through our eyes. Eyes are our

windows on the world; they tell us what is here and what is there. They
let us see clouds and rainbows, something no other sense tells us about.
We can see hair, which is less than a tenth of a millimeter thick. In fact
hair can range from 50–150 μm (μm—a micrometer—is a millionth
of a meter) in thickness, as it ranges from blond to black. Blond hair is
only half a dozen cells across, yet it is easily visible.

"Seeing" the universe, or at least "seeing" nature, is where all of science starts. We call something large or small because of the way we first
see it with our eyes or feel it with our fingers. One could not rationally
"prove" the existence of stars (except, maybe, our Sun, for we can feel
the heat) if one did not first see them. Sight is essential for our inquiry
into nature. By the end of this book we will be talking about the universe, of which we can only see 1% of 1%, and the particle universe, a
trillion times smaller than the diameter of a human hair, but we will
continue to use the language developed for our biological sensors. At
all scales things will be called large, small, heavy, light, in motion, with
energy and so forth because those are words associated with our everyday experience. Using accelerators and radio telescopes we will still talk
about "seeing nature."

So how is it that we actually see? What is that bit of ourselves that
senses the stars, the moon, the mountains, mice and even the fleas on
the hair of that rodent? The eye is an amazing organ that can sense and
distinguish things smaller than a hair and as large as galaxies. The smallest thing we can see with the unaided eye is a bit less than 10^{-4} m across;
that is, between a tenth and a few hundredths of a millimeter across.
Which leads to the question of what sets the limit of the smallest thing
you can see?

When your eyes are young you can hold an object a few centimeters
from the lens of your eye and still focus upon it. Since your eye is also
a few centimeters in diameter this means that the lens of the eye will
project an image of the object on the retina at the back of your eye,
which is just about the same size as the original object. So the limit of
the detail that you can see is set by the size of the sensors or "pixels"
of the retina. The retina is a complex network of light-sensing *rod* cells
and color-sensing *cone* cells as well as a web of nerve cells that connect
them. But in the end the size of the pixel is related to the size of a cell.
The size of a cell is about 10^{-5} m, similar to the smallest thing you can
see. Actually vision is much more complex than mere pixels because the
brain uses all sorts of information to form our internal images. Motion

and color are part of the equation. I may be able to see a single hair, but I am not so certain that I could recognize a hair if viewed end on.

What about the biggest thing you can see? In some sense, on a starry night you can see the whole universe. There is nothing between you and the most distant astronomical object, but you would not notice it. What is the most faint object that you can perceive? Here I am making the distinction between the faintest thing that fires a neuron in your retina and that which registers awareness in your brain and consciousness. The eye is amazingly sensitive; it is sensitive enough to see a single photon. The most sensitive cameras—those with charge coupled devices, which are used in the most deep-seeing telescopes—can do no better than that. But telescopes like the Hubble are patient. If a handful of photons arrive from a distant star both the eye and the telescope will sense them. If they all arrive in a brief moment the human brain will notice them. But if they are spread out across an hour our brains will be distracted and will *not* notice them. The telescope is patient and waits and stares at a single point in the sky.

So our eyes are sensitive enough to see even a galaxy beyond our own Milky Way. The naked eye can see the Andromeda galaxy, two million light years, or 2×10^{22} m away. Some people can even see the Triangulum galaxy, nearly three million light years away. But we cannot tell with just the naked eye that these are collections of stars, or 'island universes" as Immanuel Kant described them in 1755. To the unaided eye they appear as single points of dim light. It takes the optics of the telescope to gather enough light and to spread out that image and separate the stars. The unaided eye can distinguish objects that are about $0.02°$ to $0.03°$ apart. One cell of the retina, as viewed from the eye's lens, subtends about $0.02°$ to $0.03°$. So the most distant object we can see, like the smallest, is determined by the size of a cell. We can see things that vary in size from just under 10^{-4} m to around 10^{22} m (maybe really more like $\sim 10^{20}$ m), which is astonishing! So the range of vision is limited and determined by the size of a cell, which leaves us with the question, "What sets the size of a living cell?"

***

Cells are the smallest things that "live," and all living things are made of cells. What sets their size is the fact that they must somehow accomplish "life." Life is not easy to simply define. But we do know what is alive when we see it. So instead of defining life, what we tend to do is list the attributes that all living things share. If you crack open an introductory biology textbook, someplace in Chapter 1 will be a list of the attributes of life, which will include: regulates itself, uses energy, grows, adapts to its environment, responds to stimuli and reproduces. There is one other attribute that I have never seen listed. If living things grow and reproduce continuously the Earth would become crowded— unless living things also must die. When living things die it is very different from what happens when inanimate objects stop functioning. You can "kill" a computer program or turn off your car, and then turn both of them on later without any detrimental effects. But when a living thing is turned off, it stays off. Why?

Living things stay dead because they decay and fall apart and they are too complex to reassemble. They stay dead because of their complexity. How complex and therefore how large will the smallest living thing be? Erwin Schrödinger (1887–1961) tried to address this problem in his public lectures at Trinity College in Dublin in 1944. Schrödinger was a physicist and tried to cast the problem in terms that he could deal with. He said that atoms are fickle things that tend to bounce around, but that life required something with a bit more stability. The only way to make something stable out of erratic atoms was to use enough atoms so that they are "statistically" stable; that is, so that the "average atom" is sitting still. Therefore, he reasoned, a cell needs to be made up of at least tens of thousands of atoms. Schrödinger did not really pin down the size of the smallest living thing, but he asked an important question and tried to answer it. How small can a living thing be?

We now know that molecules are intrinsically more stable than Schrödinger believed and that it would not require statistical stability to give persistence to biological molecules. What seems to set the lower limit on the size of living things is that being alive is a complex problem that requires machinery of some sophistication. Living things must be complex enough to do things like grow, reproduce and use energy, and must be stable enough to do it repeatedly. A fire-cracker can use a great deal of energy, but only once. It takes a much more complex device, like an engine, to do it repeatedly with control and regulation.

Thousands of atoms across is the size of some very important organic molecules essential to life. Proteins and DNA can contain millions of atoms, but are coiled up into bundles a thousand atoms in length. A virus can be a thousand atoms wide. But a virus by itself does not qualify as being a truly living thing because it cannot do those things that we listed for living things, or at least not by itself. A virus can reproduce, use energy, and so forth only after it invades a real living cell. It can then usurp the machinery of the fully functioning cell and subject the cell's metabolic tools to its DNA or RNA control. A thousand atoms across is one ten millionth of a meter. It is big enough and complex enough to be biologically important, but it is not complex enough to be fully alive.

The smallest truly living things are *prokaryotes*, which are cells that do not contain internal structure. The dominant type of prokaryote is the bacteria and these measure about a micrometer, or a millionth of a meter, on a side. The famous bacterium, *E. coli*, is about two micrometers across. Bacteria are so common that it is estimated that there are about 5×10^{30} (five nonillion) bacteria on Earth. Even though they are so tiny, they still add up to tons of bacteria. They could cover the whole Earth in a layer 5 mm thick or could be gathered into a cube over 10 km on a side! Does this tell us more about the amount of bacteria on our planet, or about the vast surface area of the Earth?

***

Bigger than bacteria are *eukaryotes* cells, which are cells with internal structure. All multicellular creatures, such as flatworms, fleas, humans and trees are made of eukaryote cells. Often when people talk about cells they are referring to only eukaryote cells, and we will use the word in this way from this point on.

A curious thing about cells is that they have a very limited range of sizes. Only a factor of two separates the largest and the smallest of cells, except for truly oddball cells. The ostrich egg is technically a single cell, and the largest cell of modern times, although some dinosaurs had bigger eggs. But eggs are really cells that have been stretched around a lot of stored food for the embryo. Some nerve cells (neurons) can also be of extreme dimensions. Humans have neurons that stretch from their toes to their spine and these can be over a meter long. Giraffe have neurons several meters long running the length of their necks. Squid have neurons a millimeter thick and several centimeters long. But when we are talking about these extreme lengths, we are talking about the *dendrites* of the nerve cells, which are really just curious appendages to these cells. The body of the cell itself still tends to be about ten micrometers across. The human red blood cell is often cited as a typical cell, at just under ten micrometers across. This is, as you will recall, about one tenth the thickness of a human hair, or about the thickness of common cooking foil. But why is it that essentially all cells, from the feathers of falcons to the root hairs of fir trees, are about the same size?

The primary factor that sets the size of cells is the mechanism for getting nutrients into, and waste out of, a cell. Nutrients are absorbed through the surface of the cell. If a cells grows and doubles its length it will also increase its surface area by a factor of four and its volume by a factor of eight. This is such an important point in the size of biological systems that it is worth repeating. If I have a square that is 1 m on a side and I double the length scales, all lengths will double. The length of a side has gone from 1 m to 2 m. The perimeter will increase from 4 m to 8 m, because it too is a length. But the area goes up by a factor of four, or "doubled times doubled," from 1 m² to 4 m². In three dimensions, volume grows even faster than surface area. So as a cell grows, the volume of the cell grows *faster*, much faster, than the surface area (see Figure 2.1). Volume and surface are not "scaling." This relationship between surface and volume will prove to be one of the most important concepts in setting the size of cells. Not only cells, but whole animals, the nucleus of an atom and the size of stars, are also determined by the surface-to-volume ratios.

In cells, when the length doubles the volume increases by a factor of eight. The amount of nutrients needed also increases by eight. But the

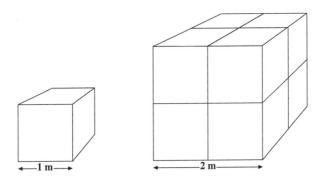

Figure 2.1 The relationship between surface and volume. If you double the length of an object, the surface area increases by a factor of four and the volume increases by a factor of eight. This is true independent of the shape of the object.

surface area only increases by a factor of four. The surface is where nutrients are absorbed. It is the entrance point, and so the cell cannot easily increase the amount of food it takes in to satisfy the bulging appetite of the greater volume.

This problem of balancing surface to volume also affects macroscopic things like towns and cities. Cities need railways or expressways to balance their "surface"-to-"volume" ratio just like living things. Imagine a square town that is 1 km on a side and is home to 10,000 people. This square town has a boundary of 4 km, and radiating out of it are four roads—maybe a north, south, east and west road. Enough food and other supplies come in on these four roads to feed the population, and enough waste is trucked out to keep the community healthy. So this means that each road carries the food and waste for 2500 people, and the traffic level is perfectly acceptable.

Now imagine that this town has grown into a city and its population has grown by a factor of 100, but it has the same population density of 10,000 people per km^2 (see Figure 2.2). The city is now 10 km on a side, it covers 100 km^2 and it has a population of 1,000,000 people. It is still square, but now it has a boundary 40 km long and so it may have 40 roads upon which supplies are brought in and waste is removed. But each road now serves 25,000 people. There will be ten times as much traffic on the city supply roads as on the town's supply roads! Also the original part of the city, the old town, which is presumably some place

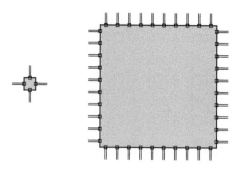

Figure 2.2 As towns grow into cities the traffic for supplies must increase. The town (left) has 10,000 people and 4 roads and so 2500 people per supply road. The city (right) has 1,000,000 people and 40 roads and so 25,000 people per supply road.

near the center of the city, needs to have all of its supplies and waste moved through the small streets of the surrounding neighborhoods. The solution? The city will build arterial roads, which are dedicated to carrying more traffic. When Baron Haussmann reshaped Paris in the middle of the nineteenth century he may have been motivated to build great avenues and boulevards to make the city more pleasant—as well as to control the mobs—but he also provided traffic pressure relief. Railroad corridors, great roads and highways allow a city to grow. But between the avenues and the boulevards, you could still have village-like streets; that is, unless you started to build taller buildings and had an even greater population density!

There is one other important factor that determines cell size; living things are supposed to control and regulate themselves. The mechanism for regulation is that proteins are produced in the cell's nucleus and they diffuse out into the main body of the cell. When a cell doubles in length and the volume increases by a factor of eight, the nucleus needs to work eight times harder and produce eight times as many proteins to control all that volume. But it is even a bit more complex than that because the proteins move by diffusion and the time it takes something to diffuse goes as the square of the distance that it needs to travel. Thus by doubling the length of a cell, the nucleus needs to produce many many more proteins to regulate itself than otherwise!

***

Now all of life is not just about cells. In fact, seeing cells as the basic constituent of living creatures is a relatively modern view. Life has traditionally been viewed in the form of multicellular organisms. These differ from unicellular creatures in that cells differentiate and take on specialized tasks, but still work together for common survival. The simplest multicellular organism is the sponge, where different cells specialize in such duties as digestion, fluid flow, structure and reproduction. But sponges lack a true body design.

Rotifers are a class of creatures that are a bit bigger than single cells. They can be as little as 100 μm (10^{-4} m), or ten cells, long. They can be made up of as few as a thousand cells in total. These aqueous creatures are as long and as wide as an average human hair is thick.

At a slightly larger scale, a few ten thousandths of a meter in size, there are thousands, if not millions of species. The flatworm *Monogenea* can be as little as a half a millimeter (500 μm) across. The house dust mite is 400 μm long and 300 μm wide. All of these tiny creatures have real structure. Structure and specialization is another way of solving the problems of nutrient flow, regulation and doing what needs to be done to be alive. Remember, if a cell grows to be too big it essentially suffocates or starves because its volume, and so its demands for nutrients, grow faster than its surface area.

So why can a collection of cells that we call an organism do much better? It has to do with the fact that cells can specialize to do one task very well when they do not have to do all the tasks required to stay alive. For example, a cell in a lung can be thin and allow oxygen in if it does not have to also be thick and provide protection. Specialization of tasks is what allows railroads and highways to move enough goods to support a dense city population. But these are not the roads that people will shop, work and live along. Likewise, specialization is why a production line in a factory can be so efficient and produce more widgets per worker per hour than a solo artisan or craftsman in their isolated studio or workshop.

However, specialization by itself does not solve the problem of living. Suppose I have a simple flea that is 2 mm on a side and is made up of ten million cells, 2% of them on the surface. This flea cannot get enough air through its skin alone, so if this were the only way it could breathe it would suffocate. But in multicellular organisms some cells can be rigid and offer structure. With rigid structure nature has devised a simple solution for breathing. The sides of an insect are perforated with *trachea*, small channels that allow air to get deep within an insect.

This solution is not unique to insects; plants also need to breathe. Plants do most of their work in their leaves, which clearly have a lot of surface area, but not enough. It has been estimated that an orange tree with 2,000 leaves has a surface area of about 200 m^2. But the air passages within the leaves, the *stomata*, effectively increase the surface area by a factor of thirty to 60,000 m^2 or 6 hectares (15 acres).

Larger than this are slugs, worms, snails and various insects. As the body forms grow in variety it becomes harder and harder to assign a single "length" to the whole species. As an example of the problem of quantifying the size of a species with a single number let us look at *Littorina saxatilis*, the smooth periwinkle. This is a snail that lives in the intertidal zone on the rocky coastline of the Atlantic. Although at first glance the periwinkle would appear to be a nearly spherical creature,

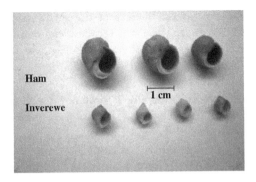

Figure 2.3 A collection of *Littorina saxatilis* (periwinkle) collected in Scotland. Courtesy of D. C. Smith.

defining its size is not so easy. In fact this is a problem that will plague and pursue us across the universe. What is the size of a periwinkle?

A recent collection of nearly a thousand periwinkles from across northern Scotland gives us a nice data set, some of which are shown in Figure 2.3. After measuring all these shells it was determined that the length of the spiral of *Littorina saxatilis* is 10.5 ± 2.2 mm. But in that simple statement we have lost a lot of other interesting information. We only know the length of the spiral—just one number. The smallest periwinkle in the collection was 5 mm long. Is this as small as they get, or is it just a limit due to the method of collection? Their size would clearly vary with age. It also varies with the environment. In the town of Ham the average shell was about 14 mm long where as in Inverewe it was only 9.5 mm. What number should we quote in the end? For our purposes it is enough to say that a periwinkle is bigger than a dust mite or flea, yet smaller than a pygmy shrew. It is about 1 cm in size.

***

Just a bit larger than the periwinkle, at the scale of a few centimeters, a lot of new structures appear within animals. For instance, the passive trachea, which served insects so well, is replaced with lungs and gills. These respiratory systems are active in that they move air through an animal and, with the aid of blood and a circulatory system, they can get oxygen deep within a bigger creature.

A few centimeters is also the size of the smallest mammal, the Etruscan pygmy shrew and the smallest bird, the bumblebee hummingbird, both weighing only a few grams. The Etruscan pygmy shrew is also the smallest warm-blooded animal. The term warm-blooded is out of favor among biologists who instead use the term *homeothermic*, which means that maintaining a constant internal temperature. Smaller animals are cold-blooded or *poikilothermic*, because they have such a high body surface-to-volume ratio that they cannot keep themselves warm. So once again here is a boundary set by the surface-to-volume ratio. An organism below the size of the Etruscan pygmy shrew has so much surface area for its small volume that it will cool and heat quickly and its body temperature will follow that of the air or water where it lives.

Scale really does affect what can happen. There is this idea, popular in fiction, that we could have fully functioning humans of extreme sizes, like the Lilliputs from *Gulliver's Travels,* "not six inches high" (15 cm), or the Brobdingnag, the giants from that same book. Better yet, science fiction likes mutant beasts such as Godzilla or King Kong. However, in reality these beings just would not work. One of the first people to realize this was Galileo Galilei (1564–1642) who wrote about the effects of scale in his book, *The Discourse and Mathematical Demonstrations Relating to Two New Sciences,* often simply called *Two New Sciences.*

Galileo starts out with a series of observations. Why can we build a small boat with no supports, yet at the arsenal shipyard of Venice, large ships need stocks and scaffolding to hold them together while under construction or they will rip themselves apart? Or, as he noted, "a small dog could probably carry on his back two or three dogs of his own size; but I believe that a horse could not carry even one of his own size." A small stone obelisk or column can be stood up on end without difficulty, but a great obelisk like the one in St. Peter's Square or Cleopatra's Needle will go to pieces when lifted unless a special cradle and scaffolding are used to stand it up on end. Cats can fall distances without harm that would injure a man or kill a horse. Finally, he observed,

Nature can not produce a horse as large as twenty ordinary horses or a giant ten times taller than an ordinary man unless by miracle or by greatly altering the proportions of his limbs and especially his bones, which would have to be considerably enlarged over the ordinary.

I was told once that an elephant the size of a cathedral would crush its own legs. It is a curious calculation that I could not resist following through. The largest elephant is *Loxodonta african*, the savanna elephant. The largest savanna elephant ever measured was shot in Angola in 1956. It stood 4.2 m at the shoulders and weighted 12,000 kg, but this was an extreme case. An average savanna elephant is about 3 m tall at the shoulders and weighs about 6000 kg. Let us scale this up to a cathedral. The main vault inside of Notre Dame de Paris is 34 m high, whereas York Minster is 31 m. Let us scale our 3-m elephant to 30 m at the shoulders. If its height increases by a factor of 10, then its length and width would also increase by a factor of 10, so its volume and mass would increase by a factor of 1000. That means that our *Loxodonta cathedralus* will weigh 6,000,000 kg (see Figure 2.4). How thick will its leg bones need to be to hold this up? Curiously enough, four bone pillars of 30-cm diameter will suffice, for bone really is tough under compression. Bone has seven times the compression strength of concrete. But an animal is different because it can move, and a single step of this 6,000,000-kg elephant will shatter a 30-cm diameter leg bone.

If we simply scaled all the length dimensions of a real elephant the bones would have one hundred times the cross section, but they would need to carry one thousand times the mass, so they are not really thick enough for an active elephant. However, all we really need to do is

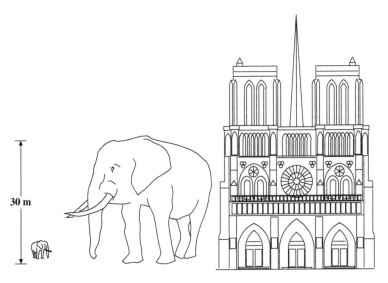

Figure 2.4 Savanna elephant *Loxodonta african* (3–4 m), *Loxodonta cathedralus* (30 m) and Notre Dame de Paris (34 m).

increase the strength of the bones by a factor of ten, which means increasing the cross section by an additional factor of ten. If we do this it should be about strong enough to walk, except . . . a normal elephant is 13% bone by mass. Our reinforced *Loxodonta cathedralus* will have 10 times as much bone, ten times that 13%, which is a real problem. So in the end, if we scale our elephant to the size of a cathedral it can stand there because bones are strong under compression. But either we increase the percentage of bone, until it is only a skeleton, or as soon as it takes a step, it will break a leg. Super-monstrous creatures are not going to happen in nature.

***

So how big can an animal grow to be? The largest creature to ever walk the face of the earth was the *Argentinosaurus*. It has been estimated that it weighed 80,000–100,000 kg. Other dinosaurs were taller (the 18-m *Sauroposeidon*) or longer (the 40-m *Supersaurus*), but none was more massive. The *Argentinosaurus* was about eight or nine times the mass of the largest modern elephant (12,000 kg). That means it is like a savanna elephant with the lengths scaled up by a factor of two. These prehistoric giants were only a factor of two longer, taller and wider than our modern elephants. And like their modern counterparts, they were vegetarians!

Why is it that the great predators, the top of the food chain, the ultimate hunters and killers are not the biggest creatures out there? The *Tyrannosaurus rex* (~10,000 kg) was dwarfed by the *Brachiosaurus* (~50,000 kg) and the *Argentinosaurus*. The largest tiger (*Panthera tigris*; ~280 kg) and the lion (*Panthera leo*; ~250 kg) are much smaller than the elephant. Even the largest terrestrial hunters, the polar bear (*Ursus maritimus*; ~500 kg), is less than half the size of the elephant. The reason has to do with the energy density—the calories per hectare of land—available in the biological world.

***

The largest creature ever to exist does not worry about crushing its legs—it does not have any. The largest creature ever seen on Earth is alive today and swimming in the oceans: the great blue whale (*Balaenoptera musculus*). Longer and sleeker looking than most whales, the blue whale measures over 30 m from the tip of its flat, U-shaped head to the end of its great tail fluke. The longest whale ever measured was 33.6 m long. The heaviest was 177 tons (177,000 kg). These leviathans have solved the structural problem of crushed legs and ever-growing bones by floating and not trying to support their mass on four brittle pedestals. The blue whale is not a vegetarian, but has solved the "calories per square kilometer" problem by eating krill, a type of tiny shrimp, which is plentiful in the ocean. This may in fact be the largest difference between the size of the predator (30 m) and the prey (0.01 m), and one of the shortest of food chains: phytoplankton (10^{-5} m) to krill to whale.

**

But why limit our discussion of living beings to only the animal kingdom? If we are looking for the largest living thing we must also consider flora, the kingdom of plants. The tallest tree on earth is the California redwood, *Sequoia sempervirens*, a member of the cypress family. For many years it was thought that the tallest was the Stratosphere Giant, measured at 112.8 m and living in Humboldt Redwood State Park in northern California. However, recently (2006) a new tallest tree, the 115.5-m Hyperion, has been found in nearby Redwood National Park. I like the idea that there may still be undiscovered giants someplace in those forests. The largest redwood by volume is the Del Norte Titan, with an estimated volume of 1,045 m^3. A number of studies predict that under ideal conditions a redwood may grow to 120–130 m in height. The limiting factor is that the tree cannot move water up any higher. But it takes 2,000 years to grow a tree to these heights, and it is unlikely for conditions to be ideal for such a long period.

I have stood in Humboldt Redwood State Park at the bases of some of these trees. Before arriving I remember thinking how odd it is to name trees in the same way we name people. However, these are not just ordinary trees growing in the forest. These giant trees define the space and allow the other trees and plants of the forest into their presence. The towers of York Minster are only 60 m tall, about half the height of the Hyperion. The towers of the Brooklyn Bridge are just over 80 m above the water.

**

While the redwoods may be the tallest trees, they are not the biggest. The giant sequoia *Sequoiadendron giganteum* owns that title. Although the largest sequoia, the General Sherman, is only 84 m tall, it is estimated to contain 1,500 cubic meters of wood, nearly 50% more than the Del Norte Titan. These trees, measuring over 30 m around the trunk, are truly giant. Their soft, cork-like bark seems to muffle any sound these stately groves may have. The giant sequoia grow on the shoulders of the Sierra Nevada mountains in the eastern part of central California. It is not as moist there as in the redwood groves near the coast, but the conditions must be right for longevity. These trees are not only large because genetics and design allow them to be, but also because they survive. They are all marked with scorches from forest fires, but they keep growing. The oldest giant sequoia that was cut down and had its rings counted was 3200 years old, which means it had lived since the time when the Greeks took to their thousand ships and sailed to Troy. There seems to be no fungus or fire that will kill them, they have no natural predators. But one physical constraint remains. Their normal fate is to succumb to gravity and fall under their own weight.

For two or three thousand years they have grown in these quiet cathedral-like groves. Yet "cathedral-like" is wrong. They tower over, and are more ancient than the stone handiwork of man. Perhaps instead one should say that cathedrals have towering heights like a redwood and massive, ageless presence like a sequoia.

***

There are other living things that some argue qualify as the largest on Earth. For example, all the trees of an aspen grove can be part of the same organism, for each tree can be a sprout off a single underground root system. There is also a giant fungus, *Armillaria ostoyae*, which is measured in square kilometers. But these are clonal colonies. To me, they do not qualify as a single living entity because they do not reproduce by growing another grove, and their parts are not mutually dependent. Finally, when I wrote down the characteristics of life I added the ability to die. If you cut a grove or a fungus in half it does not die; each part does not depend on the rest.

***

Bacteria to sequoia is an amazingly wide range of sizes for life to span. From a micrometer to a hundred meters is a size factor of one hundred million. Likewise, from the lifetime of a bacteria at 20 min (10^3 s) to the lifetime of ancient trees at nearly 5000 years (10^{11} s) is also a factor of one hundred million.

If you stand in a grove of giant sequoias and pick up a handful of dirt you can be assured that with a microscope you would see bacteria. In the pond water next to you you will find amoebas and rotifers. There are worms and mites underfoot, flies, bees and birds in the air. Measuring in at 0.5–1.7 m you will see *Homo sapiens*—humans—wandering slowly in awe. There are white fir and sugar pines growing there, as well as the dominant sequoia. All of these can live together because they each occupy their own unique niche, their own place in the environment. Usually a niche is described as a constraint of the environment. We recognize two separate niches because this place is wet and that place is dry, this area is warm and that region is cold, and certain plants and animals occupy certain niches. But we can also think of a niche, not in terms of the constraints or problems for living, but in terms of solutions to those problems. The problems of life may span over eight orders of magnitude, but the solutions do not. Trachea can get oxygen into an insect, whales need lungs and a circulation system. A nervous system works in an elephant, but not in a bacterium. The solutions have ranges given by the laws of nature, and it is the many combinations of these solutions that make each and every living species unique.

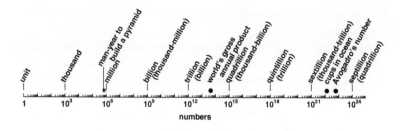

numbers

❧ 3 ❧

Big Numbers; Avogadro's Number

When I was a boy I was once in a quiz bowl contest. Our team was asked the question "Why is it easier to become a billionaire in America than in Britain?" My hand shot up and I answered that it was because a British pound sterling was worth more than an American dollar, and so it was harder to acquire a billion of them. It was 1974 and I was an American living in Britain and I knew that the exchange rate was about £1 to $2.40. But that was not the answer that the quiz master was looking for. He rephrased the question, "Why is a billion pounds in Britain more than a billion pounds in America?" None of us got the question. If it was not the exchange rate, what could it be?

What I had failed to realize is that the word "billion" meant differ-ent things on opposite shores of the Atlantic. "English" is not simply English, and we are reminded that, "America and England are two na-tions divided by a common language." I think that if the word billion had been written in French, or better yet in Chinese, I might have guessed that the confusion lay in the word itself, for there we had a very simple English word, a noun, just the name of a number. What could have been more precise than a number?

I can fill a ship in New York harbor with 1,000,000,000 nails, write on the ship's manifest "one billion nails," and send them across the ocean to London. There, the nail counter and customs officer can count out "1,000,000,000 nails" or "one thousand million nails" and, unless we know something about the translation of numbers, we may have a falling out. Is it a breach of contract to order a billion nails, but to receive only a thousand million, and think that you ordered a thousand times as many nails as you received?

The crux of the problem is that there are two conventions in the world for naming large numbers: the long scale and the short scale. The long scale is the older of these two and dates back about 500 years to Jehan Adam and Nicolas Chuquet, who wrote large numbers something like

$$1, 234567, 890123, 456789, 012345 .$$

In the *long scale* the second group of digits from the right (456789) are millions, the third group (890123) are billions, the fourth group (234567) are trillions and the left most digit (1) is a quadrillion. The convention for naming these different parts of a large number is straight forward: bi-, tri-, quad- are all prefixes that get stuck onto the front of -llion, but -llion itself is without meaning, except as invented for "million." The word million is a curiosity dating back only six or seven hundred years, and means a "great mille," or a great thousand. (Remember that a mile is a thousand steps.) The prefixes continue beyond our example to quin-, sex-, sept-, oct-, nef-, and so forth, following the Latin for five, six, seven, eight and nine (see Table 3.1). So let us apply this to a big number like the length of a light year expressed in meters,

$$1 \text{ light year} = 9460\,730472\,580800 \text{ m}$$

Table 3.1 Names of numbers.

Number	Long scale name	Short scale name	Scientific notation
1	unit	unit	10^0
10	ten	ten	10^1
100	hundred	hundred	10^2
1,000	thousand	thousand	10^3
1,000,000	million	million	10^6
1,000,000,000	thousand-million	billion	10^9
1,000,000,000,000	billion	trillion	10^{12}
1,000,000,000,000,000	thousand-billion	quadrillion	10^{15}
1,000,000,000,000,000,000	trillion	quintillion	10^{18}

This figure can be written in words as nine thousand four hundred sixty billion, seven hundred thirty thousand four hundred seventy-two million, five hundred eighty thousand eight hundred meters or, more fully, nine thousand-billion four hundred sixty billion, seven hundred thirty thousand-million, four hundred seventy-two million, five hundred eighty thousand, eight hundred meters. The words thousand-billion and thousand-million need to be there to make the long scale work. I wonder if it was this repeated use of the word thousand that eventually led to the short scale.

The short scale was also invented in France, but much later, in the eighteenth century. At this time American commerce and culture was being established, and this is the system presently used in the US. In the short scale we would read

$$1 \text{ light year} = 9,460,730,472,580,800 \text{ m}$$

as nine quadrillion, four hundred sixty trillion, seven hundred thirty billion, four hundred seventy-two million, five hundred eighty thousand, eight hundred meters. Is it a better system? Well, you only need to deal with three digits at a time, but you need a lot more prefixes.

So what system does France, the birth place of both systems use? It originally used the long form. In the eighteenth century it used the short form, but then later reverted to the long form. Actually, I expect that most people never even thought about these sorts of numbers. Even kings and the national treasury would be satisfied with millions as a big enough number.

So if I meet my old quiz master, how should I answer the question, "Why is it easier to become a billionaire in America than in Britain?" Well, I should now answer just the way I did then. "It is easier in America because one pound is presently worth more than one dollar." In fact, in 1974 the same year that this question was asked of me, the British prime minister, Harold Wilson, announced that when government official used the word billion, then it meant 1 with nine zeros. America has evolved as the world's banking center, so the way to write a national debt tends to follow the American usage or the short scale, at least in the English speaking world. An alternative is to just write out the number. For example, a few years ago the world's gross annual product was estimated to be,

$$\$44,000,000,000,000.00$$

or in a slightly different form,

$$\$44\ 000\ 000\ 000\ 000,00$$

which reminds us that we need to be careful with our commas, spaces and periods as we move into a global marketplace.

***

We can use big numbers to count things like the number of nails in a shipment, or pennies in the world's economy, or the stars in the sky. This may be useful, but it is not the most interesting aspect of a big number. What is interesting is when we can take a bit of information in the form of a number and combine it with other information to see something new and enlightening. For example, I remember reading a few years ago about a soccer game that was leading up to the World Cup. The Chinese national team was playing and work in China essentially came to a stop during the game. My question is how much work was it that was not done? There are about 1.3 billion (10^9) people in China. If half of them are working and they stop for a two-hour soccer match, that is the equivalent to 650,000 man-years of labor that was not done. I just combined the number of people, the hours for the game and the hours worked per year.

In a different place I read that some archaeologists and engineers have estimated that the Egyptian pyramids took 30–40,000 people about 20 years to build. That is, it took 600,000 to 800,000 man-years of labor to build the pyramids. That means that one soccer game caused the equivalent of one pyramid to not be built. Although soccer matches and pyramids are both interesting, what I mean to show here is that we will often want to combine these large numbers to create a new insight. I can understand an invisible thing, such as the labor force of a nation or the effect of a large disruption, in terms of the creation of a very visible object like the pyramids.

There is a classic problem that Erwin Schrödinger attributed to Lord Kelvin, which involves some pretty big numbers and which leads to some profound insights about our world. Take a cup of wine and pour it into the ocean. Now wait until the ocean waters have had time to be stirred together, time until some of the wine has seeped into both the Atlantic and the Pacific. Wait until it has diffused into the Mediterranean, under the ice caps of the Arctic and into the warm waters of the Bay of Bengal. Now dip out a cup of water from any ocean or sea on Earth. How many molecules from that original wine glass are in my sample cup?

I have seen this problem used to advance many arguments and points of view such as "Can we tell water molecules apart?," "Do the deep waters of the ocean ever really mix?" and so forth. But for us this problem is not about quantum identity or mixing; we really just want to look at the number of molecules of wine or water in a cup and the number of cups in the ocean.

We are actually going to work out Lord Kelvin's problem. Usually I will not write out the details of a calculations, but I will this time, so as to show how numbers flow through a problem. We will solve Lord Kelvin's problem by first estimating the number of cups in the ocean, using the size of the Earth, the fraction of the Earth covered by water, and the average depth of the oceans.

The Earth is about 40,000 km around, which means it has a surface area of a bit more than 510 million km^2. About three-quarters of the surface is covered by water, which is about 380 million km^2. Finally the average depth of the ocean is about 3.5 km, so the total volume of water is 1.3 billion km^3.

Now each kilometer is a thousand meters on a side, and each cubic meter contains 1000 l, and each liter can fill ten wine glasses. So here is our big number,

$$\left(1,300,000,000 \text{ km}^3\right) \times \left(1,000,000,000 \frac{\text{m}^3}{\text{km}^3}\right) \times \left(1,000 \frac{\text{l}}{\text{m}^3}\right)$$
$$\times \left(10 \frac{\text{wine glasses}}{\text{l}}\right) = 13,000,000,000,000,000,000,000$$

wine glasses of water on Earth

That is thirteen sextillion (short scale) or thirteen thousand trillion (long scale). There must be a better way of writing numbers like this because we have not even started writing out the number of molecules.

***

So how do we write out really big numbers? We can write out all the digits and take half a line or we can use words like sextillion, thousand-trillion or the more obscure trilliard, but that would send us to our dictionaries to figure out the number of zeros. Instead, we commonly use an alternative: scientific notation. In scientific notation we would write the number of wine glasses of water on Earth as 1.3×10^{22}. This convention builds on the idea that $100 = 10 \times 10 = 10^2$. The exponent—in the previous sentence the 2—tells us how many tens were multiplied together and how many zeros are in the full number. So I can write a million as $1,000,000 = 10 \times 10 \times 10 \times 10 \times 10 \times 10 = 10^6$. Now I can try a really big number like a sextillion. Since a sextillion has 21 zeros, we can write it as 10^{21}. Then the number of cups in the ocean, thirteen sextillion is

$$13 \times 10^{21} = 1.3 \times 10^{22}$$

The last way of writing it is called the normalized form.

What about numbers that are not so simple as twelve and a sextillion, such as the speed of light?

$$\begin{aligned} \text{speed of light} &= 299,700,000 \text{ m/s} \\ &= 2.997 \times 100,000,000 \text{ m/s} \\ &= 2.997 \times 10^8 \text{ m/s} \end{aligned}$$

Not only is scientific notation compact, but it can also tell is something about the character of a number. 2.997 in the above example tells us that we know this number to four digits of accuracy. Actually this is a bad example, because as we will see later, we really know this number very well.

For all of its usefulness, scientific notation is a relatively new way of writing down a number. It dates back only to the 1950s and appeared first in computer manuals. In computer notation the number of cups in the ocean would look like $1.3E+22$ and the speed of light would be written as $2.997E+08$. If you looked at scientific papers a century ago all the digits of a numbers were usually written out, but the people building the first computers wanted a simple way of storing a number in memory that was independent of how big the number was. It is also easy to do things like multiplying numbers represented by scientific notation. For example, we can calculate one of the numbers mentioned at the beginning of this chapter, the number of meters in a light-year. The number of seconds in a year is about 3.15×10^7, and so we can combine this with the speed of light to get;

$$(1 \text{ yr}) \times (3.15 \times 10^7 \text{ s/yr}) \times (2.99 \times 10^8 \text{ m/s}) = 9.42 \times 10^{15} \text{ m}$$
$$= 1 \text{ light year} = 1 \text{ ly}$$

What made this simple was that I did not multiply seven-digit numbers by eight-digit numbers. Instead I multiplied $3.15 \times 2.99 = 9.42$ and added the exponents: $7 + 8 = 15$.

So now we have a notation with which we can express really big numbers. Let us try to answer the question, "How many molecules are in the ocean?" To do this we are going to need something called Avogadro's number, and when we write down this number we will really need scientific notation.

***

Amedeo Avogadro (1776–1856) did not know how many molecules there are in a glass of wine, but if asked he would probably tell us that the number was very large but not infinite. In the development of the modern atomic and molecular description of matter Avogadro comes near the beginning, but not quite at the advent. That distinction goes to John Dalton (1766–1844), an English scientist whose interest lay not only in chemistry, but also in meteorology and color blindness, a condition with which he was afflicted. He was born in Cumbria, in northern England, he eventually moved to the bustling industrial city of Manchester where he became secretary of the *Manchester Literary and Philosophical Society*. It was at their meetings and in their publications where Dalton's atomic theory unfolded. Much of Dalton's theory remains intact even today. Matter is made of atoms, atoms come in different types, one type for each element, and atoms combine in unique ways to make molecules. It may seem obvious to us today, but Dalton was unique in that he was able to look at the chemical data of his day and see that atoms are real.

One of the key piece of evidence for Dalton is something called the law of definite proportions. For example, if I mix one part oxygen with two parts hydrogen it would all combine into water vapor. If I had used more hydrogen I would get the same amount of water vapor, as well as some residual hydrogen that did not react. It is the fact that there is this left-over residual that tells us about the atomic nature of matter.

If the law of definite proportions also applied to baking it would mean that when I combined sugar and flour to make cookies there would be only one correct ratio of the ingredients. If I had put in too much sugar, after I took the cookie dough out of the bowl, the extra sugar would still be laying there. This does not happen because cooking is generally about mixing and not about chemistry. Too much sugar and I just get a very sweet cookie.

Back to chemistry. When I say "one part oxygen with two parts hydrogen," what do I mean by the word "part"? When I am doling out the oxygen am I to measure its mass or volume, or something else?

Dalton said that for each element each part had a certain mass or weight. So for example, one part of hydrogen would be 1 g, whereas one part of oxygen was 8 g. With this technique he could explain a great number of reactions, but not everything. In fact it is this atomic weight that distinguishes his theory from that of the Greeks.

Most of the best chemical data of that time came from Joseph Louis Gay-Lussac (1778–1850), a professor at the Sorbonne, in Paris. Gay-Lussac studied various properties of gases, and even rode a hot air balloon up 6 km to study the chemistry of the atmosphere. But it was his data in the hands of Dalton and Avogadro that has had a lasting impact. Avogadro, after reading Gay-Lussac's *Memoir*, proposed that "part" meant volume and, although we rarely think of it in those terms today, he was in fact right.

So how is it that Dalton and Avogadro, two bright and gifted scientists, can both look at Gay-Lussac's data and come to very different conclusions? Their debate spanned a wide range of reactions that Gay-Lussac had measured, but we can understand the crux of the problem by looking at plain and simple water. Gay-Lussac reported that 2 l of hydrogen and 1 l of oxygen yielded 2 l of water vapor. So here is the problem from John Dalton's point of view. The mass of 2 l of hydrogen is 0.18 g, and 1 l of oxygen is 1.4 g. Their ratio is about 1:8, just as Dalton's atomic weights said. So according to Dalton, there are the same number of hydrogen molecules as oxygen molecules in these two samples.

Alternatively, according to Avogadro's hypothesis, "the number of integral molecules in any gas is always the same for equal volumes." So (according to Dalton) 2 hydrogen plus 1 oxygen should lead by Avogadro's hypothesis to 1 l of water vapor. But instead it leads to 2 l. Dalton saw this as evidence to support his view. Still, Avogadro maintained that the distance between molecules is the same in all gases, and so the number of molecules in a liter was constant.

We now know that Avogadro was right: the distance between molecules of any gas, at the same temperature and pressure, is a constant. Given a liter of oxygen, or hydrogen, or water vapor, or ammonia, or anything else there will be the same number of molecules in each and every liter.

But how do we reconcile that with Dalton's objections and Gay-Lussac's data? The crux of the problem is that Avogadro talked about molecules and Dalton talked about atoms. If I have a liter bottle full of hydrogen, it is molecular hydrogen, or H_2. Molecular hydrogen means that two hydrogen atoms are bound together. In fact they are bound close to each other and as far as their gas properties are concerned they act as a single object. They travel, bounce, mix and diffuse as one. They determine pressure and temperature and are the objects that Avogadro was talking about. So if we look at Gay-Lussac's data, we now understand that 1 l of hydrogen contains twice as many atoms as molecules. This is true of oxygen as well, but not water. Water is singular.

$$2\,l\,H_2 = 4\,l\,H_1$$
$$1\,l\,O_2 = 2\,l\,O_1$$

$$2\,l\,H_2O = 2\,l\,H_2O$$

All the reactions in Gay-Lussac's data make sense if we understand that some gases (not all) naturally occur as two-atom molecules. So what about Dalton's view of the data? Dalton assumed that simple gases (hydrogen, oxygen and so forth) were always atomic. So he explained the data by seeing water as HO! In the end, for a given volume of gas there is a set number of molecules, independent of the gas type. That is Avogadro's hypothesis, but not Avogadro's number. We still do not know how many molecules there are in a liter of a gas, or a gram of a solid, or a cup of the ocean. Loschmidt would be the first to measure the size of a molecule, half a century after Avogadro and Dalton, but to do that he needed to know something about the way molecules bounced around in a gas.

***

The study of gases was a hot topic throughout the nineteenth century, but it was nearly 50 years later before we hit the next critical step in figuring out how many molecules there are in a cup of wine. James Clark Maxwell (1831–1879), who later would become famous for his unification of electricity, magnetism and light, was studying the diffusion and mixing of gases. He was working in the 1860s and theories and experiment were in disagreement. So he developed a new model for the distribution of motion in a gas, the kinetic theory of gases. In his description, a gas is a collection of molecules that are in motion, moving around and colliding with each other. The hotter the gas, the more they move. The more they collide, the higher the pressure. At one point Maxwell wrote

If we suppose $\sqrt{k} = 930$ feet per second for air at $60°$, and therefore the mean velocity $v = 1505$ feet per second, then the value l, the mean distance traveled by a particle between consecutive collisions $= \frac{1}{447\,000}$th of an inch, and each particle makes 8,077,200,000 collisions per second.

I have included this whole paragraph because of the way the numbers are reported is interesting. First the 8 billion (short scale) or 8 thousand million (long scale) collisions per second is reported with all of its digits. Maxwell also seems to indicate five digits of accuracy, a precision that the raw data did not justify. He also reports the distance as a fraction, instead of the decimal form we presently would expect. But really it is this distance, the distance molecules travel between collisions, $l = \frac{1}{447,000}$ in $= 5.7 \times 10^{-8}$ m, that will help us solve the wine glass problem.

***

The next step came from Johann Josef Loschmidt (1821–1895), a scientist from Vienna who used Maxwell's results to give us our first really good estimate of the size of a molecule and the magnitude of Avogadro's number. Loschmidt had read all about the kinetic theory of gasses and he knew all about Maxwell's results and that the distance that a molecule travels between collisions is $\frac{1}{447,000}$ of an inch. With this Loschmidt set out to calculate the number of molecules in a given volume of gas.

Loschmidt's reasoning went something like this. The amount of elbowroom that a molecule needs is the volume of a cylinder whose cross sectional area is the same as the cross-section of the molecule, and whose length is the distance a molecule travels before it hits

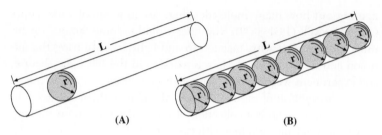

Figure 3.1 Loschmidt's method for determining the size of a molecule. (A) This cylinder represents the average space a molecule occupies, where r is the radius of the cylinder and the molecule, and L is the distance traveled between collisions. (B) When the gas is condensed many molecules fit in the same volume. We can see that $L = 2rC$, where C is the *condensation factor*. From this, r, the size of the molecule, can be determined.

something (see Figure 3.1). If there was more space than this, then a molecule would travel further before it collided. The next thing that Loschmidt needed to know was how much of a gas volume is occupied by molecules, and how much is empty. He called this the *condensation factor*, and measured it by looking at the difference in volume between a gram of liquid water and water vapor. Typically the condensation factor is about a thousand. From that he showed that the size of a molecule is the ratio of the collision distance and the condensation factor. In other words, since the condensation factor told us that only one part in a thousand of a gas is not void, then the size of the molecule is about one thousandth of the collision length. Now actually Loschmidt's equation is a bit more complex because he treats a molecule as a sphere and corrects for how tightly they can be packed to make a solid or liquid, but those are small corrections. In the end he reports that the size of a molecule is "0.000 000 969 mm" (9.69×10^{-10} m). He was measuring air, which is dominated by N_2 (nitrogen), and so by modern measurements was only off by a factor of three. But he had the right magnitude. Also it was not his reasoning, but rather the experimental numbers that were supplied to him that account for most of the differences. He described his results as follows.

An imposing string of numbers such as our calculations yield, especially when taken into three dimensions, means that it is not too much to say that they are the true residue of the expectations created when microscopists have stood at the edge of the bottomless precipice and described them as "infinitesimally small."

I think Loschmidt must have been a bit awed by his results. It is not too often that a scientist will wax so poetically, especially when writing in a journal for his colleagues to read. But he really did have a result that was breathtaking. When he stood at the edge of that precipice, it was not that he saw that it had an infinite or unknown bottom, but rather that he was the first to get a glimpse of the bottom and to know that it was down there, nine or ten orders of magnitude smaller than a meter.

Loschmidt was so fascinated by this result that he spent part of that 1865 paper trying to give the readers a feel for what this small size meant. He explained that if all the atoms in a cubic millimeter were spread out to one layer thick, they would cover 1 m^2 (actually 10 m^2). He also reported that the wavelength of light is about 500 times longer than these molecules (actually about 1500 times). He also talked in terms of the weight of an atom:

Incidentally, the calculation suggests for the "atomic weight" of chemists a suitable unit of a trillionth of a milligram.

This suggestion of Loschmidt was interesting for three reasons. First off, he was suggesting something like what we now call an atomic mass unit (u). Today we define an atomic mass unit as one twelfth of the mass of a carbon-12 atom, which is about 1.66×10^{-21} mg. The second point of interest was that Loschmidt proposed a unit that was strictly metric. His number was 1.00×10^{-21} mg. It is curious that he would suggest that the basic unit be defined in terms of the macroscopic gram, instead of the microscopic atom. The choice of an atomic mass unit in terms of carbon ended up being a practical solution. It is easier to measure the ratio of the mass of a molecule to a carbon atom than the mass of that molecule to a macroscopic gram.

Finally, is a trillionth of a milligram really 10^{-21} mg? A trillion has twelve zeros, not twenty-one. Actually it is the same because Loschmidt was using the long scale for his numbers. He was writing in Austria where a billion is a million million, and a trillion is a million billion, with eighteen digits.

★★★

Finally we can turn to one of the really big numbers in science, Avogadro's number. Avogadro made the statement that equal volumes of different types of gas contained the same number of molecules. But we do not define his number in terms of volume. Avogadro's number is a number we use for counting out stuff. It is like using the word "dozen" to mean you counted out twelve, or a "gross" to mean one hundred and forty-four (a dozen dozen). If you count out an Avogadro number of atoms or molecules you have a *mole* of that stuff. Now one mole of any gas under normal conditions does occupy the same volume. One mole of gas occupies 22.4 l, or about 6 gallon jugs. But gas can be heated such that it expands, or chilled to a liquid, or even frozen to a solid, and still a mole will contain the same number of atoms or molecules. So whereas we recognize Avogadro was right at a set temperature and pressure (one mole is 22.4 l at $0°C$ and at a normal atmospheric pressure), we now define a mole and Avogadro's number in terms of mass. Twelve grams of carbon-12 is exactly (by current definitions) one mole of that substance, and contains exactly Avogadro's number of carbon atoms.

What about a mole of water? This will be important in answering our wine glass question, since most of wine is made up of water. Water is chemically H_2O, two moles of atomic hydrogen weigh about 2 g, whereas one mole of atomic oxygen weighs about 16 g. So a mole of water is about 18 g, or we can say that our wine glass holds about 5.6 moles of water.

But how many atoms in a mole? Loschmidt told us the mass of an atom, so we can use his number to find that there are about 10^{24} atoms in our 12-g chunk of carbon. That is essentially Avogadro's number.

Actually over the last century we have improved upon Loschmidt's results a lot. What we do is we take a 12-g chunk of carbon and count the atoms in it. Well, maybe not directly. For example, the Hope diamond is about 9 g, and so three-quarters of a mole. The number of atoms in it is a 24-digit number. What we can do, with a high degree of precision, is measure the space between atoms in a crystal. The measurement of structure and spacing is called *crystallography*, and we will discuss it in more detail in a later chapter on atomic scales. For now, however, what is important to us is that we can shine an X-ray on a crystal and something like a Moiré pattern of light and dark spots is formed. From measuring these patterns we can determine the spacing of atoms in the crystal very well. But knowing the spacing of carbon atoms in a diamond crystal is not quite the same as knowing how many atoms there are. Counting the number of atoms in a crystal from spacing is a lot like calculating the number of people in a crowded room. If there is about a meter between people at a party, and the room has an area of 40 m^2, then there are about 40 people there. We would actually do a much better job of our crowd estimation if people would cooperate and stand in a regular formation. Perhaps instead of studying the party, we could try counting the people in a marching band who are all lined up in rank and file formation. If they are spaced a meter apart, an arm's length to your neighbor's shoulder is very close to a meter, and if we know the area the band covers, we know the number of people very well. This regularity is why we turn to crystals of carbon—diamonds—instead of chunks of coal. Diamonds have a regularity that repeats, row after row, column after column, layer after layer, throughout the whole macroscopic crystal.

So now, at long last we come to the famous Avogadro's number:

$$N_A = 6.02214129 \times 10^{23}$$

molecules or atoms in a mole.

Avogadro never saw this number, but based upon Loschmidt's reaction to his own results I think Avogadro would have been astonished at its size. Twelve grams of carbon contain 6×10^{23} atoms. The number was named after Avogadro more than a century after he stated his "same number of gas molecules in the same volume" hypothesis. In fact in some places, especially German-speaking countries, the number is referred to as Loschmidt's number, but I think this is becoming less common.

Now let us turn to our wine glass and ocean problem one last time. There are 5.6 moles of water in 1 dl, or 1 wine glass. That means that there are about 3.3×10^{24} molecules in our wine glass. From our previous calculation we find that there are about 1.3×10^{22} glasses of water in all of the oceans. To answer Lord Kelvin's original question, "If I pour a glass of wine in the ocean and stir the ocean completely, how many of the original wine molecules will be in a cup of water I scoop out of any sea?" The answer is the ratio of the number of molecules in a cup to the number of cups in the ocean,

$$\frac{3.3 \times 10^{24} \text{ molecules in a cup}}{1.3 \times 10^{22} \text{ cups in all oceans}} \approx 250.$$

There will be about 250 molecules of the original wine in every cup scooped out of any ocean, or any other water anywhere on Earth. This result is amazing. It not only means we can we measure atoms and calculate their number, but it also says that each and every cup is full of history. Given time, the oceans do stir themselves up, and water molecules tend to be stable across thousands and millions of years. That means that my cup of water, dipped out of the Connecticut River contains some of Lord Kelvin's wine. It also contains some of the wine used to launch Cleopatra's barge, as well as any other historic event you can think of. It contains some of the Lake Erie water Governor Dewitt Clinton poured into New York harbor to mark the opening of the Erie Canal in 1825. It also contains some of the wine Homer sipped as he composed the phrase "wine dark sea."

There are about 8×10^{45} molecules of water on Earth. Later we will look at how many atoms and even quarks there are in the universe, and maybe we will start to understand what "a drop in the ocean" really means.

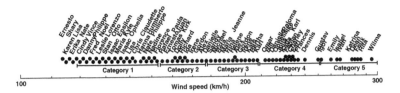

Category 1   Category 2   Category 3   Category 4   Category 5

100   Wind speed (km/h)   200   300

❦ 4 ❦

Scales of Nature

Imagine sitting in a wind that is roaring along at scores of kilometers per hour, yet not having a hair blown out of place. This may seem absurd, but that is the experience of high altitude balloonist. When one crosses the Atlantic in a gondola suspended beneath a balloon, the winds at high altitude may rush you along but you can set papers on a table and not worry about them being blown away. This is because you, the air, the gondola, the balloon and the papers are all moving together.

Measuring wind speed turns out to be a bit more tricky than it might first appear, because we cannot directly see the wind. But once we realize that balloons move with the wind, we have a handle to tackle this problem. So now before we go and measure the universe I would like to pause and talk about how we measure some simple things, like wind speed, starlight and the hardness of rocks. Actually it is not the measurement techniques that interest us here, but rather the scale upon which we plot these measurements. These scales are called *logarithmic*, but before we get side tracked on what that means, let us return to those balloons.

***

A balloon moves with the wind once it has been untethered from the Earth, and tracking a balloon is relatively easy. So by measuring the velocity of the balloon we can measure the velocity of the wind. Even today, NASA will release balloons before a rocket launch and the

National Weather Service will launch dozens of balloons daily. Not only can you track a balloon and measure wind speed, but the balloon tends to keep rising and so you can measure the speed at various altitudes. But quite honestly, it is not the simplest technique, and it tends to use up a lot of balloons. An alternative method is to use a device called an *anemometer*.

You have probably seen an anemometer. It is a simple wheel with cups designed to catch the wind and spin around. Typically the wind catchers are three or four hemispheres mounted on spokes around a vertical axis. The open side of the cups catch the wind and are pushed back. The other side of the cups are smooth and offer little resistance to the air, and so can be pushed into the wind, letting the wheel spin around. By measuring the rotation rate of the wheel one can measure the wind speed. This simple device was invented by Dr. John Thomas Romney Robinson of the Armagh Observatory in Northern Ireland in the 1840s. It is now such a common thing that it can be seen in schools and at airports. It is used at both professional and amateur weather stations everywhere. It is ubiquitous, but it is not the only way to measure wind speed, and the need to measure wind preceded the anemometer.

Historically people would describe the wind in terms like a "stiff breeze," or a "moderate wind," or a "gale." But the description of wind was inconsistent until Sir Francis Beaufort of the British Royal Navy established what we now call the Beaufort scale. In 1805–6, Beaufort was commanding HMS Woolwich, conducting a hydrographic survey off of the Río de la Plata in South America. He first developed his scale as both a shorthand notation for the wind conditions, as well as a set of standards to make his reports consistent. For example, if he had just enough wind to steer by, he called it a "light air," and marked it as a 1 on his scale. If he had enough wind so he could sail at five or six knots he called it a "moderate breeze," and recorded it as a 4.

His system might have remained a private shorthand, but a sniper wound cut his sea career short and led him to a land-locked position at the British Admiralty. Here, he was eventually appointed Hydrographer to the Admiralty, and by 1838 the Beaufort wind force scale was the standard way that all ships in the Royal Navy recorded wind conditions.

The Beaufort scale has evolved over time, but it has always been based on the idea that the number reported is derived from an effect that can be seen. Originally all the standards were things seen on sailing ships. If a ship needed to double reef its sails, the wind was a Beaufort 7.

But ships evolved away from sails and sailing conditions, and so the scale was adopted to the conditions of the sea. A Beaufort 7 is called a "moderate gale," and on the ocean we see that "the seas heap up and foam begins to streak." On land you can recognize a 7 because "whole trees are in motion and an effort is needed to walk against the wind." What is missing from our twenty-first century point of view is an objective, absolute scale that would somehow allow us to relate Beaufort's scale to kilometers per hour. But before that let us look at extreme winds.

An extension of the Beaufort scale is the *Saffir–Simpson hurricane scale.* This is the scale that is used when we hear on the news that "Hurricane Katrina has been upgraded to a category 5," or "Sandy is a category 1." This scale is similar to the Beaufort scale in that it is based upon the expected effects of the storm. In a category-2 hurricane, one can expect the loss of roofs and damage to mobile homes and poorly constructed structures. In a category 5, such as Katrina in 2005, most trees are blown over and most manmade structures can expect major damage.

In about 1970, Herbert Saffir, a structural engineer, was studying hurricane damage for the UN. He developed the scale to help prepare for the aftermath of storms. If a category 3 is coming ashore, relief agencies will already know approximately how much damage to expect and how many resources they should be ready to deploy. Later Bob Simpson added the effects of the storm surge to the damage calculation, which makes a lot of sense, since hurricanes that are formed over the seas do most of their damage when they make their landfall. Actually, since 2010 the scale has been redefined to not include the storm surges, the reason cited being that these effects are hard to calculate and are often very local.

A curious trait of both the Beaufort and Saffir–Simpson scales is that these scales are primarily about the effects of the wind and not the wind speed. But the two measurements can be related. I can hold up my hand held anemometer and measure the wind speed under various Beaufort conditions. The translation has now been standardized as:

$$v = (0.836[m/s])B^{3/2}$$

Here v is the velocity of the wind and B is the number on the Beaufort scale. This equation is worth dissecting because so much of nature has a similar form. The "0.836 [m/s]" means that we are translating from the Beaufort scale to a velocity measure in meters per second. If we wanted miles per hour or kilometers per hour there would have been a different constant here. It is the 3/2 that is really interesting. It tells us that if the Beaufort number goes up by four, the velocity has gone up by eight. The velocity changes faster than the Beaufort number. If I double the wind speed, the effect is less than doubled. As the equation is written above we would say that the "velocity rises exponentially with the Beaufort number." We could turn around the equation, and write the effect as a function of the wind speed as:

$$B = \log_{3/2}(v)$$

which tells us that the physical effect (B) rises logarithmically with velocity (v) (see Figure 4.1).

The fact that the effects that are felt change more slowly than the quantities that we measure pervades all aspects of nature. In Chapters 1 and 2 there was a graph that related the size of a wide range of objects such as atoms, cells, planets and stars by plotting them on a logarithmic scale. But when you looked at those plots you do not see the word

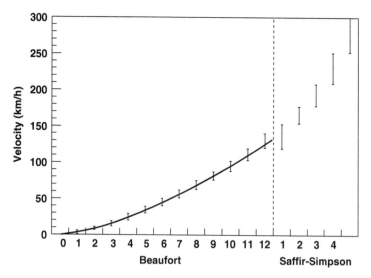

Figure 4.1 The Beaufort and Saffir–Simpson scales and windspeed. The Beaufort and Saffir–Simpson scales are a quantification of the effects of the wind. We see that the resulting effects rise slower then the actual wind speed.

logarithm anywhere. What you do see is that for every step to the right the sizes get bigger by a factor of ten and every step to the left they get smaller by that same factor. We could have written 1, 10, 100, 1000, 10000 and so forth, but instead we wrote 10^0, 10^1, 10^2, 10^3, 10^4, using scientific notation. Now every increase is just a step of one in the exponent. For us it was a convenient way to relate the size of neutrons, whales and galaxies on the same graph. But there is more to it than just convenience. Nature really is this way. Phenomena change more slowly than sizes, distances, time, or energy.

***

Another scale that measures the destructive effect of a natural phenomenon is the Richter scale of earthquakes. Us humans are always wanting to compare things and so a very natural question is "was this earthquake I just felt bigger than the one someone else felt a year ago?" Before I build a "bigness"-measuring device I had really best decide what that means. I could argue that since an earthquake is all about motion, energy has been released, and therefore I would like to compare the *energies* of different earthquakes. But measuring that energy is not so simple and in the end it might not be the best way to describe an earthquake as it relates to humans.

Much like the Beaufort scale, the Richter scale arose out of a technique of measurement. Charles Richter and Beno Gutenberg were researchers at the Carnegie Institute and Caltech where, in 1935, they were using a Wood–Anderson seismograph to measure earthquakes. This was a standard device at the time. In it a paper chart slowly rolled past a suspended pen. When the ground shook, the pen would quiver and sway and leave a squiggle on the paper. Now the problem facing people trying to interpret seismic data was that different reporting stations would measure different size squiggles for the same quake. The reason was obvious: some stations were nearer to the epicenter, the source of the earthquake, than others. Richter developed the *local magnitude scale*—commonly called the Richter scale—as a method of adjusting for this. Richter combined the distance from the epicenter and the measurement of the seismograph to determine the magnitude of the earthquake with his equation:

$$M_L = \log(A) + 3\log(d) - 2.92$$

Here A is the amplitude of the squiggle measured on the chart and d is the distance from the epicenter. What this means is that if you are a standard distance from the epicenter (100 km) a magnitude 1, 2, or 3 earthquake will set your pen swaying by 1, 10, or 100 micrometers. A magnitude 5 will set the pen swaying by 10^5 micrometers, which is 10 cm or 4 inches.

The Richter scale has been supplanted by other seismic scales for a working geologist, but it still thrives in the public consciousness. It is still a scale we talk about and it is the way that the magnitude of a quake is reported in the news. So how big is a magnitude 1 or 4 or 8? A magnitude 1 earthquake can be caused by 32 kg of TNT, a typical blast at a construction site. A magnitude 2 could be caused by a ton of TNT, which is a large blast at a quarry. Going from 32 kg to a 1000 kg in one step of this seismic scale, we can see that the energy—the amount of TNT—which is needed to cause an earthquake one step greater is rising much faster than the Richter magnitude. The great earthquake of San Francisco of 1906 was a magnitude 8.0. The largest earthquake ever recorded, the Chilean earthquake of 1960, was a 9.5.

As a side note, another interesting seismic scale is the *Mercalli intensity scale*, named after Giuseppe Mercalli (1850–1914). Mercalli was a priest and professor of the natural sciences who studied volcanology in Italy. His scale was based on the effects felt by people. For example, a Mercalli intensity II would be felt by people, especially people in the upper floors of a building. It is equivalent to a Richter 2 or 3. An intensity VII would cause some damage in well-built buildings and major damage to poorly built ones, and is equivalent to a Richter 6. What is so interesting to me about the Mercalli scale is that it is based purely upon the experience of people near the quake. Moreover, the Mercalli scale tracks almost exactly with the Richter scale (see Figure 4.2), which is based upon the squiggles of a shaking pen or, more precisely, upon the logarithm of the size of the squiggles of that pen or the logarithm of the energy of the quake.

Finally, we can use the magnitude of the quake, either on the Richter scale or the Mercalli scale, to estimate the resulting damage. This is a lot like the way we use the Saffir–Simpson hurricane scale. The US Geological Survey tells us that magnitude 3 earthquakes are "often felt, but rarely cause damage," whereas a magnitude 8 "can cause serious damage in areas several hundred miles across."

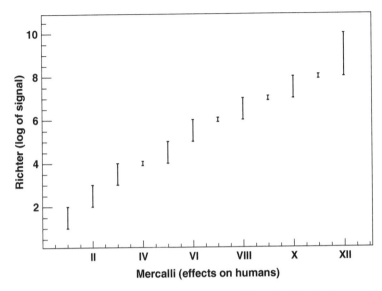

Figure 4.2 The Richter scale versus the Mercalli scale. The Richter scale (based on the logarithm of the pen motion, or energy) tracks with the Mercalli scale, which is based on human experience.

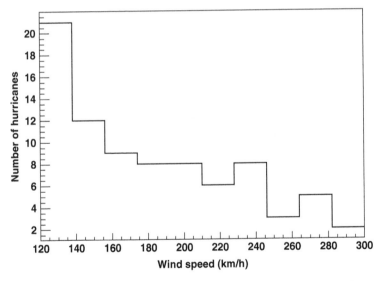

Figure 4.3 The number of hurricanes in the North Atlantic (2003–2012) as a function of windspeed.

One other similarity between earthquakes and hurricanes is that as the severity of the destruction increases, the frequency of the occurrence decreases (see Figure 4.3). Thank goodness!

***

While on the subject of geology, one other curious scale worth mentioning is the Mohs scale of mineral hardness. This is the scale where diamonds, the hardest natural substance, is a 10 and talc is a 1. Friedrich Mohs (1773–1839) developed his scale of mineral hardness in about 1812. What made Mohs' scale so useful is its simplicity. He started with ten readily available minerals and arranged them in order of hardness. How can you tell which of two stones is harder? You simply try to scratch one with the other, and the hardest stone always wins. Quartz cannot scratch topaz, because topaz is harder. On the Mohs scale quartz is a 7 and topaz is an 8. Since the ten standards are relatively common, a geologist heading into the field needs to only carry a pocket full of standards to determine the hardness of a new, unknown sample with a scratch test.

Is it the best, most objective and quantifiable scale? No, not really. Ideally hardness is a measure of how much pressure is needed to make a standard scratch. Conversely we could drag a diamond point with a standard pressure across the face of our sample and then measure the size of the scratch. This is the technique behind a *sclerometer*. We can use a sclerometer to make an absolute measurement of hardness. Not only do we know that quartz (Mohs 7) is harder than talc (Mohs 1), but we now know that it is 100 times harder. A diamond, that perfect lattice of carbon atoms that makes nature's hardest substance, is a Mohs 10. More than that, it is 1500 times harder than talc or 15 times harder than quartz.

The second hardest mineral on Mohs' scale is corundum. This oxide of aluminum (Al_2O_3) comes to us in several forms including rubies, sapphires and emery. Emery, of course, is that grit found on emery cloth, which is used to "sand" steel. Since it can scratch nearly everything, it is not surprising to find it high up on the hardness scale. At the other end is talc, which we usually encounter as a powder and find hard to imagine as a stone. Soapstone is primarily talc, valued because its softness allows it to be easily carved.

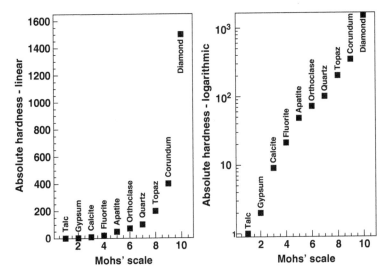

Figure 4.4 The Mohs scale versus the absolute hardness of minerals. The Mohs scale was designed to accommodate a simple technique, but in fact is close to the logarithm of absolute hardness.

What makes the Mohs scale so interesting for us is that as you go up the scale, from talc to gypsum, calcite, fluorite and apatite, the hardness just about doubles with each step. The ascent in hardness continues through feldspar, quartz, topaz, corundum and finally to diamond. If hardness were to increase by 2, or even 2.2, with each step, the scale would clearly be logarithmic. It does not rise with perfectly tuned steps (see Figure 4.4)—the step between gypsum (2) and calcite (3) is a factor of 4.5 on an absolute hardness scale—but it is close. It is not logarithmic because Mohs' objective was to create a simple test that could be used in the field. The fact that it is almost logarithmic is accidental, except that nature seems to like logarithmic scales. On a practical side there are a great many more soft rocks than hard rocks, so Mohs partitioned his scale where the field is most dense. Finally the Mohs scale is based on standard collection of rocks and not on an ideal; it is based on the technique of comparison and not an absolute scale.

***

One other scale that was originally based upon comparing samples to standards is the star magnitude or brightness scale. Curiously enough it was the word "magnitude" for stars that inspired Richter to use the same word for earthquakes.

Our earliest encounter with stellar magnitude is in Claudius Ptolemy's catalog of stars from about 140 AD. Ptolemy was part of the intellectual Greek society in Alexandria, Egypt, at the time of the Roman Empire. It was a community known for its library and built upon the great Greek academic tradition. Ptolemy's *Almagest*, with the position and magnitude of over a thousand stars, is the oldest catalog of stellar brightness that we still have, but it is based upon the lost works of Hipparchus (∼190–120 BC) who ranked stars into six different magnitudes of brightness. In this system the brightest stars are first magnitude stars. Figure 4.5 is based on data from the *Almagest*. Stars that are half as bright as first magnitude are second magnitude. Stars that are half as bright as second magnitude, or a quarter of the brightness of a first magnitude, are labeled third magnitude. And so on down to sixth magnitude stars, which are about the dimmest things you can see with the unaided eye. So this scale, with the difference between magnitude being a jump of two in brightness, is a logarithmic scale.

I think it is worth our effort to understand stellar magnitude because later in this book the stellar magnitude scale will be one of the major tools used for determining the distances to stars, and therefore the size of the universe.

Ideally the magnitude of a star, as determined by Hipparchus or recorded in *Almagest*, is determined by just looking at the star. In practice it is hard to judge brightness and half-brightness; the naked eye does not determine these things precisely. If every step really was a factor of two, then the sixth magnitude star would have one thirty-second of the light of a first magnitude star or 3% of the brightness. But by modern measurements of stars *Almagest*-rated sixth magnitude stars have about 1% of the light of a first magnitude star. The eye is actually more sensitive than Hipparchus or Ptolemy appreciated.

In the *Almagest* the stars Rigel, Sirius and Vega were all rated first magnitude, even though their intensity, by modern measurements, varies by a factor of over four. This seems to suggest that the Greeks did not know how to rate the stars. In fact the technique that they used was to recognize a few standards and then compare new stars to nearby standards. For example, if Ptolemy called Polaris, the north star, a 3, then

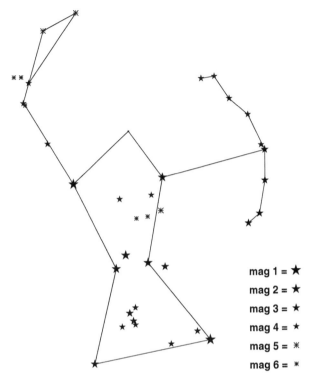

Figure 4.5 The stars of Orion as reported in the *Almagest*.

mag 1 = ★
mag 2 = ★
mag 3 = ★
mag 4 = ★
mag 5 = ✳
mag 6 = ✳

you could decide that Electra, a star in the Pleiades was dimmer and so could be rated a 5. The art of rating stars is then a lot like Mohs' scale, where the new samples are compared to old and established samples.

The science of measuring starlight, or any light, is called *photometry*, and by the middle of the nineteenth century people were making much more precise measurements. This was the age of the telescope and the eye, before electronics sensors or even photographic plates. So astronomers used a clever bit of optics called a *heliometer*. The heliometer was originally designed to measure the size of the Sun's diameter as it varied over the year (as the Earth moved), but it had a use far beyond the Sun. The heliometer could take two images—two views of the sky—and with internal mirrors put them next to each other. Now the technique for measuring star brightness is to first always start with the same star: Vega, with magnitude 0, and therefore even brighter than magnitude 1. With the heliometer one could view Vega and the star one wanted to rate, say Polaris, at the same time. It is very clear that Polaris is dimmer, but by how much? What one does now is cover up part of the telescope that is pointing towards Vega until both stars appear to have the same brightness. In this case one needs to obstruct three quarters of the light from Vega to get them to match. So Vega, magnitude 0, has four times the brightness of Polaris, and therefore Polaris has a magnitude of 2 by modern measurements.

By the middle of the nineteenth century measurements like this were becoming quite good and systematic and they revealed a deep problem. Stars that the *Almagest* said were magnitude 6 were really much fainter than 3% of the brightness of a first-magnitude star. Astronomers either needed to relabel these dim objects as magnitude 8, or adopt some other way of describing brightness.

Norman Robert Pogson (1829–1891), as an assistant at the Radcliffe Observatory in Oxford noted that the first-magnitude stars are about 100 times brighter than the sixth-magnitude stars and that stars could keep there traditional magnitude designation if we allowed that the difference between magnitudes to be about $2\frac{1}{2}$ instead of 2. More precisely, we can take the factor of 100 and divided it into steps of 2.511886... = $\sqrt[5]{100}$, the fifth root of one hundred. Today we call this number the *Pogson ratio* (P_0). This also gives us a prescription for calculating magnitude based on an absolute luminosity (L), the number of photons per second:

$$m_{\text{vega}} - m_{\text{new star}} = P_0 \log \left(\frac{L_{\text{new star}}}{L_{\text{vega}}} \right)$$

This would be all well and good, if Vega were truly a star of steady brightness.

Modern photometry has moved beyond the techniques used by Pogson and his colleagues and we can essentially count photons received from a star. But why would we want to do this? We now understand that if two stars are the same color—if they have the same hue of red or blue, for example—they are probably about the same size and putting out the same amount of light. Then, if one star is much dimmer than another star of the same color, it must be farther away from us. Thus we can use star brightness to measure the distance to stars and by implication, the size of the cosmos.

So Pogson patched up Hipparchus's stellar magnitude system, making it the precise and objective scale we still used today. We now also have a few new terms. When we measure starlight we say that we are measuring a star's *apparent magnitude*; that is, the magnitude as it appears to us on Earth. But to compare stars we will want to determine a star's *absolute magnitude*; that is, how bright would that star appear if we were a standard distance from that star. This is the same thing that is done with the Richter scale. The seismologist does not report what the seismograph measures; rather, he reports what they would have read if the seismograph was 100 km from the epicenter. The standard distance for stellar absolute magnitude is 10 parsecs, or 3×10^{17} m. As an example, if you were 10 parsecs from our own Sun, it would appear to be a dim magnitude 4.83 star. However from our vantage point of Earth, 1.5×10^{11} m away, the sun is 42,000 times brighter than Vega, or an apparent magnitude of -26.74!

So why is it that Hipparchus and Ptolemy were so far off and labeled some stars as sixth magnitude when, by their own system and criteria, they should have been labeled eighth magnitude? First off, I do not think people realized how sensitive our eyes really are. The fact that we can see a star that has one percent of the light of Vega is amazing. Vega is itself is just a pinprick of light in the black void. But when we use only our eyes we are comparing the impression left on our brains. I can tell that a sixth magnitude star is dimmer than a fifth magnitude, but quantifying it as one percent of Vega is much harder.

For example, back on Earth, if I was asked the length of a backyard telescope that is about a meter long, I could estimate its length by eye to about 10 cm. However if I was asked to estimate the size of an array of radio astronomy telescopes, scores of antennas that look like radar dishes arranged in a line 10 km long, I could only guess its size to about 1 km, not to 10 cm. We can estimate things to about 10%, a relative size, not to 10 cm, an absolute size. This is the central tenet of Fechner's (or Weber–Fechner's) law. Sensation increases as the logarithm of the stimulation; that is, slower than the stimulation.

At the same time as Pogson was looking at stars, Gustav Fechner (1801–1887) was performing the first experiments in *psychophysics*. Psychophysics is the study of the relationship between a physical stimulus and how that stimulus is psychologically perceived. For example, Fechner would have a blindfolded assistant sit in a chair and hold a brick. Then he would add weights on top of the brick and ask the assistant if they felt an increase. If the addition was too little the assistant would not notice it, but if the addition exceeded some threshold, the assistant knew it. It would take a number of trials to establish this threshold, adding and subtracting weights in a random order, but eventually the threshold could be measured. Now Fechner would give the assistant a new brick of a different weight and measure a new weight perception threshold. He found that if the weight of the brick was doubled, the threshold would also double.

The actual threshold will change from person to person, for some people are more sensitive to physical stimuli than others, but the fact that the threshold is proportional to the basic brick's weight seems to be universal. Mathematically this means that the sensation is proportional to the logarithm of the stimulus. One of the consequences of this is that even though there is a fantastic range of stimuli to our eyes and ears, there is a much smaller range of sensation when these show up in our brains. For example, we can see things as bright as the Sun (magnitude -27) and as dim as the Triangulum galaxy (magnitude 6), which means that our eyes can span 33 magnitudes of stars, or in intensity $P_0^{(6-(-27))} = 2.511^{33} = 1.6 \times 10^{13}$. There is a factor of 16 trillion between the brightest and the dimmest stars we can see. This number is a little bit deceptive, since the light of Triangulum galaxy appears as a point, whereas the Sun subtends about half a degree, which means the sunlight is spread out across our retinas. Our ears, however do not spread out the stimuli, and still they have a marvelous range.

***

When it comes to sound, we measure how loud something is with the units of decibels (dB). A whisper is about 30 dB, a vacuum cleaner is about 70 dB, and a rock concert can be in the 100–130 dB range. The sound of breathing can be 10 dB, and the faintest thing you can hear if your ears are young and undamaged is near 0 dB. So our range of hearing goes from 0 to about 120 dB and, right in the middle at 50–60 dB, is normal conversation. But what is a decibel?

Sound is a pressure wave moving through the air. When you speak you push on the air near you, and this bumps up against the molecules of air a bit farther away, which bump up against the next neighbors and so forth. It is not that the air travels, but rather that the "push" travels, i.e. the pressure travels. So perhaps we should be measuring sound in terms of pressure or the energy of the waves. The reason we do not is because the decibel really is a useful unit for describing the sensation we feel in our ears. So what is a decibel?

The decibel was first developed at Bell Labs. They were originally interested, not in how loud sound is, but rather they were trying to quantify how much a signal would deteriorate as it traveled down a wire. Originally the people at Bell Labs would describe the deterioration as "1 transmission unit," or 1 TU. A TU was the amount of deterioration of a signal as it traversed 1 mile of standard wire. Again, a unit is defined in terms of a standard. In 1923 or 1924 they renamed the TU a *bel*. It has been suggested that they were trying to name it after their lab, but defenders of the name will point out that most scientific measurement units are named after people—volts, watts, amps, ohms, henrys, faradays and so forth—and that the bel is named after the father of the telephone, Alexander Graham Bell. But a bel is not a decibel.

A decibel is a measurement of the magnitude of the ratio of something to a standard. Here the word "magnitude" refers to how many factors of ten there are in this ratio. So 10, 100, 1000, or 10^1, 10^2, 10^3 are of magnitude 1, 2 and 3. Then the definition of a decibel is:

$$ dB = 10 \, \log_{10} \left(\frac{P_1}{P_0} \right) $$

Here P_1 is the power of what we are measuring and P_0 is the standard we are comparing it to. The 10 in front of the log is where the prefix "deci" in decibel comes from. At Bell Labs this meant that 10 miles of standard wire would be rated at 10 dB, 100 miles at 20 dB and 1000 miles at 30 dB. Decibel can refer to any ratio and engineers use it widely. But to most people it refers to sound and to how loud something is.

In sound, the standard was selected to be the faintest thing good ears can here, which is a pressure wave of 20 μPa; that is, a pressure of twenty billionths of an atmosphere. Normal conversation is about 60 dB or a pressure wave one million times (10^6) more powerful than that faintest sound. Our range of hearing, from 0 dB to 120 dB, is a factor of a trillion in power! So why do we talk about decibels instead of pressure or power? It is because 50–60 dB really does seem like the middle of our range of hearing, whereas a million does not seem like the middle of the range between zero and a trillion. If our hearing range goes up to a trillion, should half a trillion not be mid-range? Imagine you are 100 m away from a jet engine and hearing 120 dB, or a trillion times that faintest standard. If you move away about 40 m, the power will have dropped in half. The sound will be at 117 dB, which is still beyond comfort. So the decibel is a very practical way of describing the range of our hearing and much more useful than pressure or energy.

We could actually use the decibel to describe any ratio. For example if our standard was the light from a sixth-magnitude star, then the Sun would be rated at 130 dB.

***

There is one last scale that I will describe in this chapter: the musical scale. Actually there are many, many different musical scales and the discussion of the nuances of a certain scale can be long, technical and even heated. For now I will confine myself to the *diatonic* or *heptatonia prima* scale. This is the scale most pianos are built for and tuned to. You may have already spotted the "dia-" prefix and realized that two, or a half, will be our multiplier between different magnitudes of this scale. The "-tonic" refers to tones.

As we have just said, sound is a pressure wave that passes through the air. But it is not a single wave; it is a series of waves. Different musical notes are characterized by the different frequencies of these waves. According to the ISO (International Organization for Standardization), the A-key of a piano that is just above the middle-C should be tuned to 440 Hz. If you press down on the A-key of the keyboard a small hammer behind the keyboard is tripped and will strike the A-string. That string will vibrate 440 times per second. If you placed a microphone next to your piano and watch the electrical signal from the microphone on an oscilloscope or a computer screen, you would see the signal rise and fall 440 times each second. But the signal you see will not be a simple sine wave. A simple sine wave would mean that we were hearing a pure tone, or a *monotone*. That is the sort of sound we associate with push-button telephones. Our piano has a much richer sound and a more complex wave pattern. We can think of these complex sound waves as a combination of many simple waves. The sound from our piano's A-note is made up of a wave with frequency 440 Hz plus a wave with frequency 880 Hz, 1320 Hz, 1760 Hz and so forth. All these additional waves are multiples of that original 440 Hz of the monotone A. The 440 is called the fundamental frequency, while the 880, 1320, 1760, 2200 are the *harmonics* or the *overtones*. What gives an instrument its unique sound, or its *timbre*, is the way its combines these overtones: the relative weighting or strength of the first, second, third . . . overtones compared to the fundamental harmonic. You can hear an A from a piano and distinguish it from the A of a guitar or a trumpet or a flute because your ear takes the sound apart, breaking it into the harmonics, and your brain recognizes that when (say) the second harmonic is strong and the third and fifth are weak the A is from a certain instrument. In fact your ear is so good at sorting out all of these harmonics that you can listen to a whole orchestra and pick out individual instruments, even when they are playing the same note.

So back to our A above middle C, which is also called A4 and is the 49th key on the keyboard. Its first overtone, or second harmonic, is 880 Hz, which is also the frequency of A5, the A in the next octave up the scale. If I double the frequency, I jump up an octave. If I divide the frequency in half, I drop down an octave. This is the origin of the name *diatonic*. So for a musical scale 2 is our multiplying factor, much like the Pogson ratio for stars, in which the factor is $\sqrt[5]{100} = 2.511886$. Likewise, the word "octave" plays a similar role for music as "magnitude" does for stars.

But what about all those other notes besides A? You will recall that I said the diatonic scale is also called the *heptatonia prima*. "Hept-" is seven and tells us about the seven whole notes, or white keys, in an octave. But there are also five black keys, giving us twelve distinct notes in an octave. When we move up the keyboard from A to $A^\sharp/B^\flat$ to B to C, $C^\sharp/D^\flat$, D, $D^\sharp/E^\flat$, E, F, $F^\sharp/G^\flat$, G, $G^\sharp/A^\flat$, we increase the frequency at each step by 5.946%. This is curious: a step up the keyboard or up the scale is a jump by a percentage, and not by a set frequency. This also means that the percentage difference between two adjacent keys anywhere on the keyboard is identical. There are twelve steps in an octave, and after those dozen steps our frequency will have doubled. So the key-to-key multiplying factor is the twelfth root of two: $\sqrt[12]{2} = 1.05946$ or 1 plus 5.946%.

This curious number, 1.05946, also shows up in a more visual and beautiful place: on the fret board of a guitar (see Figure 4.6). The distance from the saddle or bridge—the point where the string of the guitar is attached to the body—to a fret determines the frequency, and so the note, that that string will make when plucked. If you now move your fingers up one fret, you are a note lower, and the active string is

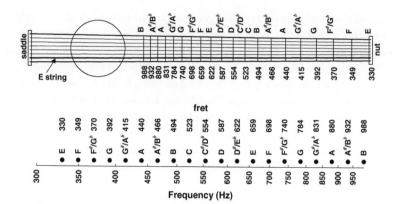

Figure 4.6 The frets and related frequencies of a guitar. The frets are laid out logarithmically. In the upper figure, each fret is labeled by the note and frequency it will produce when the E-string (bottom string) is plucked. The lower plot shows the frequencies on a logarithmic scale. Here the frets are evenly spaced.

5.946% longer. This means that the frets of a guitar are a type of logarithmic scale. When a rock star steps out onto a stage with a Fender Stratocaster, the frets under his or her fingers are laid out like a slide rule, star magnitudes and the Richter scale.

***

Somewhere near the end of this book we are going to understand how ancient and how big the universe is and how small and fleeting quarks are. So why have we just spent this whole chapter talking about wind speeds, earthquakes, star brightness and sound? It is firstly because logarithmic scales are all around us and we use them every day when we talk about loud speakers, hurricanes and pianos. But, more importantly, nature organizes itself in terms of these sorts of scales. When later we move from stellar distances to galactic distances we will not say the distances have had 10^{18} km added, but rather that distances have been multiplied by a million. When we go from cells to periwinkles to sequoias we do not add sizes, we multiply them. Nature organizes itself into different distance scales. The atomic scale is 10^{-10} m, the scale for the nucleus of an atom is 10^{-15} m and that is a whole different world with a whole different set of phenomena.

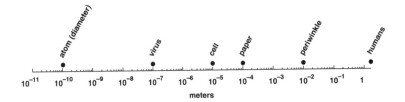

5

Little Numbers; Boltzmann's and Planck's Constants

I remember as a small child being told about air and, quite frankly, being skeptical. Here was a substance that I was told was more important for life than either food or water, yet I could not even see it. I could smell good things like lilacs and foul things like barnyards, but I could not really smell air. I could put out my hand and try to grab it, but it was too etherial and I would always come up empty handed. Air was far to elusive for me to understand.

This was in contrast with wind. Wind is much easier to understand than air. Wind lifts up kites and blows away dandelion seeds. Wind can drive a pinwheel and chase a sailboat across the water. Wind is not as invisible as air because I can see it splash through a tree, shaking its leaves and branches. I never doubted wind; it was always very real. And then I understood what air was. Wind is all about something in motion, something all around us, something that flows. Wind is that stuff that other people call air, but only when it is in motion. Therefore air must be that stuff that, when it moves, causes flags to flap, hay fields to wave, and dry leaves to skip across the lawn. Wind is air with motion and so conversely, air is wind, with or without motion. But to me as a child, wind was much more interesting than air because it was happening; it had motion and dynamics. Wind had energy.

***

Our world is intrinsically a dynamic place, always with things in motion. In fact our brains seem to be hardwired to notice motion. You can stare into the woods and not see a single animal among the thousands of branches, leaves and trunks until a chipmunk or a sparrow moves. Then our consciousness focuses on that creature. It is the animal's dynamics that grabs our attention. We are attuned to motion, dynamics and the expenditure of energy.

A ball that has been pitched and is curving over the green earth has energy. A puck sliding on ice, a planet in orbit, or an electron circling the nucleus of an atom all have energy. Energy is what makes things happen. Without energy we would (at best) be living in a still life, a place where time is meaningless, a static and silent world.

Wind is air with energy.

Energy gives the world life and dynamics, but energy is more than just motion. Energy is also the possibility or the potential of motion. A skier at the top of a snow filled slope has the potential of great motion by just pointing their skis downhill; dynamite and gasoline can cause motion with the aid of a spark; a sweet dessert ought to drive us to greater motion than it often does.

There is one more category of energy beside motion and potential: heat. Heat is a type of energy that does not fit. It is not the pent up energy of the skier or a wound up spring, and it does not appear to have much to do with motion. The glowing iron in a blacksmith's forge just sits there. But heat really is energy and it was the study of heat that led us to quantum mechanics. And in that field, in the act of *quantizing* light and energy, we will find Boltzmann's and Planck's constants and the Planck length, one of the smallest things in nature.

***

Heat is energy. This most basic statement eluded scientists for centuries. Up until the eighteenth century heat was seen as being its own substance: *caloric*. Wood that could burn contained a lot of caloric and by burning the wood you were releasing the calories. But then Count Rumford (Benjamin Thompson 1753–1814) made a most astute observation. Thompson was an American inventor who, as a loyalist, left New England during the American Revolution. He passed through London and eventually settled in Bavaria where he was attached to the army. It was at the arsenal in Munich where we start our story of heat.

At the arsenal cannons are cast, but the casting is rough and they need to have their bores drilled out. This would involve a large drill bit and a pair of horses walking around and around. At the best, it was a long process, and at the worst, with a dull bit, it produced a lot of heat. So Rumford convinced the arsenal to perform an experiment for him. He place the raw cast cannon in a barrel of water, with a very dull drill bit turning in the bore. He then started the horses turning the bit and then measured the temperature of the water. As time went on the water became hotter and hotter. It was wrapped in flannel to insulate it and keep the heat in. After two and a half hours the water started to boil! Rumford had calculated how much caloric should have been in the metal and there was not enough to boil that much water. In fact, according to the caloric theory, he should have eventually been able to drive all the calories out of the metal, at which time there would be no more heat. But that is not what he observed. As long as the horses kept walking and the drill kept turning, heat was produced and the calories were not exhausted. Rumford concluded that the energy in the motion of the drill, which was really the energy from the horses, was being converted into heat and that heat was energy and not its own separate substance.

***

The fact that energy could be transformed into heat was only one side of the coin. Nicolas Léonard Sadi Carnot (1796–1832) was a French military engineer who was studying steam engines' efficiency and wrote a book in 1824 entitled, *Reflections on the Motive Power of Fire*. In other words, he was addressing how you get motion out of heat, which is what a steam engine is doing. One of his chief discoveries is something that we still call a *Carnot cycle*. A Carnot cycle is often presented as a graph of the state of an engine in terms of heat, pressure and temperature. Various stages of a cycle are characterized by expansion of gasses, or temperature changes, or pressure falling. In a steam engine, steam is heated and the pressure rises. The piston expands and the pressure drops. The steam is condensed or released. The temperature drops and the cycle starts again.

What was revealed in Carnot's graphs and analysis was the efficiency of the engine. The efficiency told you how much fuel you had to put into an engine to get a certain amount of motion out of it. This was terrifically useful at the dawn of the age of steam, but it also contained a curiosity. It showed that there was always an inefficiency, a piece of the energy in the wood or coal that we could never quite tap into, no matter how clever we were at pressures, pistons, temperatures and condensers. Here we have seen the first hint of *entropy*, heat energy of an unusable form.

One way to picture entropy is to think of the energy in a ball propped up on a hillside. We can extract energy from the ball; that is, we can convert its potential energy into motion by tipping the ball and letting it roll down the hill to the bottom of the valley. The amount of motion is related to the differences in height between the hilltop and the valley bottom. But the ball has not given up all of its energy. If the valley flowed down into a plain or the ocean the ball could roll further and gather greater speed. Or if there was a deep mine shaft it could fall further. However, it does not. It is at the bottom of the basin and there is no place left for it to roll. Local geography gives it no further opportunity. It is at its lowest potential and we can extract no more energy from it.

Carnot's heat engine can only be perfectly efficient if the heat is infinitely hot, or the condensers work at absolute zero. Otherwise there is still energy we cannot tap into.

***

I am trying to tell a story of how our ideas of heat and thermodynamics lead us logically along a path where first we realize that heat is a type of energy, then that energy is conserved (the first law of thermodynamics), and finally that entropy always increases (the second law of thermodynamics); it would be nice if history followed that same sequence. But history has its own agenda and did not produce events in the order I would like them to have done. I would have liked to start with the most basic statement about energy, namely that energy is conserved. But that idea really was a latecomer. After Carnot it looked like some energy was lost into entropy every time it was transformed. It is only after demonstrating that heat is energy and that heat may have a very real but unusable form called *entropy*, that we can formulate the law of the conservation of energy.

This is what Rudolf Julius Emanuel Clausius (1822–1888) finally did in 1850. He welded together energy, heat and entropy in a coherent form that we now know as the first two laws of thermodynamics. First, energy is conserved. It is neither created nor destroyed, but it *can* change its form. Secondly, when it changes its form some of it may become an unusable form that is called entropy. Within a system entropy will always rise and that rise cannot be reversed.

The second law of thermodynamics was a very new and different type of law to what people were used to. It states that as time goes on entropy can only rise. This is in contrast to the great laws of physics: Newton's laws of motion. Newton's findings served as the archetypes of such laws: they showed us what physical laws should be like. Newtonian mechanics is also about moving and energy. According to Newton's laws, if billiard balls collide, energy is conserved by being transferred from one ball to another. But all the events that people used these laws for could be reversed. Let a bat hit a ball and Newton's laws tell us what to expect. If you reversed time and ran the video backwards, you would see a ball hitting a bat and still Newton's laws would tell you the outcome: Newton's laws work forward and backwards in time. Thermodynamics is very different; time flows in only one direction.

There have been a number of scientists, including the astronomer Eddington and physicist Einstein, who have said that the second law of thermodynamics is the most fundamental principle in nature. It has also been suggested that entropy is tied to the fact that time only flows into the future. Without entropy we could, in some sense, not distinguish yesterday from tomorrow. So in the second half of the nineteenth century the second law, the continuous increase of entropy, was not only seen as fundamental, but as sacred.

It would take Ludwig Boltzmann to finally link Newton's laws and the second law of thermodynamics, and he would trigger a revolution in the process. Because he dared to touch the second law he would bring an avalanche of criticism and persecution upon himself, and would be driven to the brink of insanity.

***

Back in Chapter 3 we encountered James Maxwell and his description of a collection of molecules bouncing around inside a bottle. What he was developing was his *kinetic theory of gasses*. His goal was to start with a model of molecules bouncing around, let them follow Newton's laws of motion, and try to derive such properties as pressure and temperature. Raise the temperature and the molecules have more energy; they move faster and have harder collisions with the walls of the container, which we see as an increase in pressure.

A number of other people, including Daniel Bernoulli (1700–1782) had tried to derive gas laws from Newton's equations, but Maxwell brought a great deal of mathematical prowess to the problem. For instance, instead of just describing the molecules as all having some motion, Maxwell derived the distribution of motions the molecules must have. His work was a real *tour de force*, but its sophistication may have defeated a number of his contemporaries. In Austria, however, it struck a resonance.

In the 1860s Joseph Stefan (1835–1893) was building up a physics institute at the University of Vienna. He had already attracted Loschmidt, who made the first estimate of the size of a molecule. He also captured a young student, Ludwig Eduard Boltzmann (1844–1906), and gave him a copy of Maxwell's writings and a book of English grammar to help him read it. Boltzmann completed his PhD in 1866 and in 1869 obtained the chair of mathematical physics at Graz, in the southwest corner of modern Austria. From his position there he published a paper on the kinetic theory of gasses, in which he put Maxwell's distribution on even firmer mathematical footings. In fact we now call that distribution the *Maxwell–Boltzman distribution*.

Boltzmann also derived the second law of thermodynamics from the kinetic theory of gasses. In doing so, he stepped into a hornets' nest that would dominate his reputation and career for the rest of his life. Boltzmann had a hard time publishing his description of heat because, as the journal's editor reminded him, molecules were only "hypothetical." Chemists, following Dalton, were well ahead of physicists at this time.

The key to Boltzmann's success was that instead of dealing with the motion of individual molecules he dealt with the statistical distribution of their motion. Instead of talking about a molecule traveling at "930 feet per second" (283 m/s) he would talk about the probability of there being molecules traveling at that speed. He would also talk about the evolution of the probability distribution and not the evolution of the molecule's trajectory. This was taking Maxwell's ideas of distribution and extending them to one more level of abstraction. We now call the method developed by Boltzmann *statistical mechanics*, and it concerns the dynamics and evolution of the statistics and not the dynamics of the particles.

In 1873 Boltzmann took a post at Vienna, back at the cosmopolitan capital of the country and at the center of active physics. It was here that his old friend Loschmidt explained to him the problem of statistical mechanics—what is now referred to as *Loschmidt's paradox*. Statistical mechanics is based on Newton's laws of motion, which are reversible. However, Boltzmann had derived the second law of thermodynamics—the fact that entropy always increases—which is not time reversible. The two theories therefore appeared to be logically inconsistent and incompatible. A number of Boltzmann's contemporaries felt that there must be a deep flaw in statistical mechanics to have produced this inconsistency, but it was not clear how.

To Boltzmann, a system starts with a high degree of order and decays into a state of disorder. Order and entropy are seen as being opposites. To illustrate this point, picture a pool table with fifteen balls racked up in a triangle with a game ready to start (see Figure 5.1). At this moment there is a high degree of order. Now when the cue ball breaks the racked balls, the balls scatter in all directions and roll for a few seconds. The balls are now chaotic, with a low degree of order and a high degree of entropy, even though every ball and every collision exactly followed Newton's laws of motion.

Now imagine that we could take every ball and turn it around and have each one roll back over the path they had just traced out. A ball that was headed north would now be rolling south; a ball that skidded east would now be skidding west. It would start out looking chaotic,

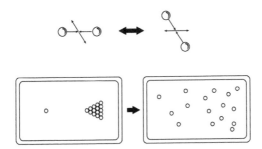

Figure 5.1 Reversible and nonreversible collisions on a pool table. Top: Newton's laws of motion describe balls colliding with each other, a process which is time reversible. Bottom: statistical mechanics describes systems that evolve from order to disorder and which are not time reversible.

but each ball would be following Newton's laws. As time continued the balls would collect at the center of the table, form a triangle, and kick the cue ball out. Newton's laws allow this, but thermodynamics does not. If you saw a video of balls rolling into a formation you would know that someone had just reversed the video.

Here is a second example. Take six pennies and a die and start with all the pennies heads up (see Figure 5.2). This is a highly ordered state. Now roll the die and when a 2 turns up turn the second penny over. Now keep rolling the die and turning pennies. At the beginning you had six heads, then five, then four and then five (you just rolled a second 2). After a while all the order is lost and you usually have three heads and three tails, but four heads is not unheard of. In the language of Boltzmann, all heads (or all tails) has the lowest entropy, because there is only one way (or *microstate*) of having this arrangement. The situation with three heads and three tails has the highest entropy because there are 20 ways of having this arrangement.

In this toy statistical mechanics model the pennies occasionally become more ordered and entropy decreases. But this occurrence in conflict with the second law. Boltzmann's statistical mechanics actually predicts that sometimes entropy will decrease. So which is right, statistical mechanics or thermodynamics?

Back in our toy experiment with pennies and a die, if I have three heads and three tails there is about a 3% chance that in three turns I will return to that orderly state of all heads. But this was only with six pennies. If instead I start with a dozen pennies with half heads and half tails it would take six turns to get to the maximally ordered state, but the probability is only about 0.024%, or 1 in 4000. On our pool table, with fifteen balls, the probability of returning to the initial ordered state is incredibly low because we are not just dealing with a two-state, heads-or-tails system, but instead we have continuous numbers to describe

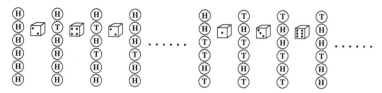

Figure 5.2 Order as shown with a die and pennies. Order generally decreases, but occasionally it can increase and entropy decrease.

positions and velocities of each ball. In the real world, where the objects are molecules and the number of them is given by Avogadro's number, the probability of returning to an ordered state is vanishingly small.

We now understand that Boltzmann's statistical mechanics is right and does form the theoretical basis of thermodynamics. In a small space, for a short time, entropy may decrease, but we will never notice it.

This is where Boltzmann's story should end. The Austrian mathematical physicist, with a growing beard and failing eyesight, should have been able to rest on his laurels. He had given us two equations and a constant that bear his name. First the Maxwell–Boltzmann distribution, which told us how energy is distributed among molecules at a certain temperature. Secondly, Boltzmann's entropy equation told us that entropy is proportional to the logarithm of the number of microstates of the same energy. Finally, he gave us Boltzmann's constant, or k_B as we often write it. Boltzmann's constant is the amount of energy you need to give to one molecule to raise its temperature by 1°C. Boltzmann's insight was that the world was really made up of a huge number of atoms and molecules and to understand heat you needed to turn to a statistical perspective of that micro-world. Boltzmann locked heat, energy and entropy together at the atomic and molecular level.

Boltzmann once wrote:

In the restaurant of the Nordwestbahnhof I consumed a leisurely meal of tender roast pork, cabbage and potatoes and drank a few glasses of beer. My memory for figures, otherwise tolerably accurate, always lets me down when I am counting beer glasses.

From: *A German Professor's Journey Into Eldorado*

Boltzmann was waiting for a train in Vienna to start his trip to Caltech in California; he also traveled to Cambridge where he and his ideas were well received. I think he should have been able to spend his later years mis-counting beer glasses in Austria, but he had one staunch nemesis in his hometown.

***

At this time Vienna was an intellectual hotbed, an academic carnival. This was the time when Sigmund Freud was opening our eyes to the mind and in Boltzmann's own department Ernst Mach (1838–1916) was challenging the way we look at the world. Mach was one of the precursors of the logical positivists and the Vienna circle of the 1920s. He called for science to confine itself to what was directly observable. This meant that Mach rejected "hypothetical" particles such as atoms and molecules since they could not be directly observed. Because of this he ridiculed Boltzmann's work and claimed it had no role in a true scientific discussion.

Mach's point of view sounds so foreign to us that it is easy to dismiss him as an eccentric. How could any sensible thinker reject Dalton's atom and the success of nearly a century of chemistry? But Mach was no lightweight and had solid physics and intellectual credentials. He was trained in Vienna under Joseph Stefan a few years before Boltzmann. He also went to Graz and then spent a number of years as a professor at Prague before being called back to Vienna. We still associate his name with the speed of sound and Einstein cited *Mach's principle* as an important inspiration for general relativity. But Mach would not accept the results of Boltzmann and the two men became locked in philosophical combat. It may have been this conflict, in part, which led Boltzmann in 1906 to take his own life.

One of the tragedies of his death was that in 1906 the tide was turning. Max Planck was describing light in terms of atom-like oscillators and Einstein was describing *Brownian motion*, the jittering of microscopic dust as the result of atomic and molecular collisions. At the time of his suicide the atomic evidence was mounting and the scientific world was ready to follow Boltzmann's lead.

***

Max Karl Ernst Ludwig Planck (1858–1947) was a very cautions and reluctant convert to Boltzmann's ideas. He had been trained in thermodynamics and even wrote his thesis on the second law, but his was a classic view. In the 1890s Planck, then at the University of Berlin, became involved in the problem of *blackbody radiation*. This problem is built upon a curious observation. When you take a piece of metal and heat it hot enough it will start to glow. If you have seen a blacksmith pulling a piece of iron out of a forge it is easy to picture the red or even white light the hot iron radiates. The color it glows in fact tells you the temperature of the metal. What is surprising is that the color is independent of the type of metal. Take a piece of steel and heat it to 900°C and it glows orange. Take a piece of copper that starts out looking very different from steel, heat it to 900°C and it too will glow with the exact same color. In fact start with anything, including an ideal blackbody with no intrinsic color, and if it is 900°C it will glow orange. 700°C will glow red and 1300°C will glow yellowish white. Gustav Kirchhoff had pointed this out in about 1860 and suggested that this observation was a hint to something deeper in nature.

In 1894 Max Planck was commissioned by the local power company to investigate this problem with the idea that the light bulb design could be optimized. Planck was not the first one to try to put an equation on the blackbody spectrum. In 1884, Joseph Stefan had published the *Stefan–Boltzmann law*, which described the energy radiated as a function of the temperature; in 1896 Wilhelm Wien published a description of the blackbody spectrum that was good, but not quite right. Also Wien's law was descriptive, but lacked a firm theoretical foundation.

Planck went back to basics. He knew thermodynamics and he knew Maxwell's equations of electromagnetic radiation, which is the foundational theory for light. He had the classical explanation of heat and light at his fingertips and he tried to weld them together, but it did not work and so Planck continued to search. He was also very aware of the work of Heinrich Hertz; Planck had been Hertz's student in Berlin shortly after Hertz did his pioneering radio work. So Planck knew that radiowaves and any other type of electromagnetic radiation, including light, can be produced when charges oscillate, like currents in an antenna. This raised the question of what were the oscillators in blackbody radiation? Maxwell's equations described light as the oscillation of electric and magnetic fields. What Planck needed was to break fields into tiny pieces in the same way that matter was made of atoms. Planck called this piece of light a quantum, what we now refer to as a photon. His final result, and equation for the intensity of light,

$$E = \frac{2ch}{\lambda^5} \frac{1}{e^{ch/k_B \lambda T} - 1}$$

was inspired by and dependent upon Boltzmann's work (see Figure 5.3). The c is the speed of light, λ is the wavelength (think color), k_B is Boltzmann's constant and T is the temperature. There remains one mysterious constant: h. We now call this *Planck's constant*.

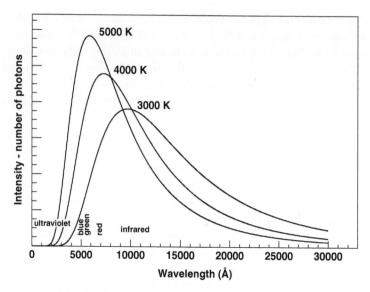

Figure 5.3 The distribution of light described by Planck's law. The distribution of light from a hot object will shift towards blue as it gets hotter. The peaks of the curves do not match the colors seen at the blacksmith because the eye does not weight all colors of light the same.

Experimental measurements were very good, much better than any theoretical prediction before Planck. They were also very extensive. Experimentalists could set the temperature and measure how much red, orange, or blue light there was. They could even measure the amount of infrared and ultraviolet. And then they could change the temperature and measure all of these intensities again. The experimentalists had dozens of numbers and Planck had one equation with two unknown constants: Planck's (h) and Boltzmann's (k_B). With all that data you could use some of it to fit these numbers and the rest to confirm them. The match was perfect! Not only were the two constants well determined, but the amplitude and shape of the spectrum, the distribution of the colors of light, was just right.

The 1901 paper in which Planck first published his equation is often cited as the beginning of quantum mechanics because it postulated that the electromagnetic fields, which gives us light, are made up of quanta or photons. The 1901 paper also contains the first appearance of h, the most basic unit of energy in an oscillator. As a side note, sometimes Planck's constant is written as $\hbar$. This is really the *reduced Planck's constant*, and is related to Planck's constant by 2π: $h = 2\pi\hbar$. Numerically:

$$h = 6.6 \times 10^{-34} \text{ J.s}$$

where J stands for joules, the basic metric unit of energy and s is seconds.

Planck's constant shows up in essentially every quantum mechanical equation, for example in *Planck's Relation:*

$$E = hf = \frac{hc}{\lambda}.$$

Here E is energy and f is the frequency of light. For instance, if I have a red laser pointer with wavelength $\lambda = 630 \text{ nm} = 6.3 \times 10^7$ m, then the energy of one photon from that pointer is $E = 5 \times 10^{-20}$ J, which does not sound like a lot. But if this is a 1 mW laser, it is producing 2×10^{16} photons each second. I could also apply this to a kitchen microwave oven. A typical oven produces microwaves at 2.45 GHz (2.45×10^9 cycles per second) and applies about 700 W of energy to the food. That works out to only 2.6×10^{-25} J per photon, but with 2.7×10^{27} photons per second.

***

There is a debate among people who study the history of science as to who was the first person to really understand what a quantum was and what quantization really meant. Some will point to Planck's 1901 paper and say, "There it is. Planck must have understood what a quantum was to be able to derive his results." In later years he would write that at the time he had seen quanta not as real and revolutionary, but rather as something that would solve a problem. I am reminded of Murray Gell-Mann, one of the originators of the quark hypothesis, who at one time referred to quarks as a convenient mathematical construct that was to be discarded after the desired results were obtained. Gell-Mann eventually saw quarks as real, and Planck eventually embraced the reality of photons and quanta.

If it was not Planck who first understood the significance of quanta, who was it? Most readers of science history agree that by 1905 Einstein appreciated that they were real. Whereas Planck used h to describe one effect—blackbody radiation—Einstein successfully applied it to an independent problem, the photoelectric effect. In truth the acceptance of quanta and quantum mechanics went through fits and starts and really was not of a form we would now recognize until the 1920s. But it really was revolutionary and one should be slow and cautious when embracing radical ideas. Planck was telling us that Maxwell's electromagnetic equations, a true Victorian *tour de force*, were not the last words on light. The world on the microscopic, or even sub-microscopic level was a bit different than what we were used to, and a bit stranger.

***

Planck's constant, $h = 6.6 \times 10^{-34}$ J.s looks like a conversion factor between energy and frequency of oscillation, but it is deeper than that. Planck's equation works because there are distinct, integer numbers of photons. The fields are not continuous at the smallest scale. That really is at the heart of the revolution and h is just the signature of these discrete bits. h, with its 10^{-34}, sure looks like a small number, but that is in part because of the units we have chosen to measure energy and time in. If we had picked electron-volts (eV), a unit useful in atomic physics, we would find that $h = 4.1 \times 10^{-15}$ eV.s, still a small number, but not quite so frightening. If we also chose an exotic nuclear particle's lifetime—the delta ($t_\Delta = 10^{-20}$ s)—as our timescale, then we would find that $h = 4.1 \times 10^5$ eVt$_\Delta$, or 410,000. So the size of Planck's constant is somewhat artificial. I would, however, like to compare it to a meter, or the size of an atom. In that case it is still very small.

At about the same time as Planck introduced his constant he also proposed a set of *natural units*, a set of length, time, mass/energy, charge and temperature that were not based on a macroscopic standards like the meter, but rather on intrinsic constants of nature. The Planck length is defined as

$$l_\text{P} = \sqrt{\frac{\hbar G}{c^3}} = \sqrt{\frac{hG}{2\pi\,c^3}} = 1.6 \times 10^{-35}\ \text{m}$$

and is illustrated in (see Figure 5.4). The length itself is defined in terms of Planck's constant, Newton's gravitational constant (G) and the speed of light, all universal constants.

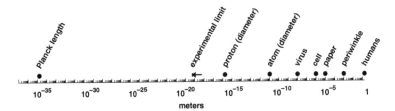

Figure 5.4 The Planck length. The Planck length is much smaller than everything else. It is 20 orders of magnitude smaller than a proton. Our biggest accelerator probes structures halfway in size between humans and the Planck length.

The significance of the Planck length is not clear even now, but there are reasons to think it may the ultimate distance standard at small scales. Nothing can interact with a particle smaller than a Planck length, as we will see in Chapter 11.

At the beginning of this book I said that the smallest scale we could probe, the present experimental limit, was a bit less than 10^{-18} m. These measurements involve the energy of our largest accelerators. The Planck length is over a quadrillion times smaller, so I do not expect direct experiments on the Planck scale in the near future. Only time will tell us what is the smallest, most tiny, final iota of nature.

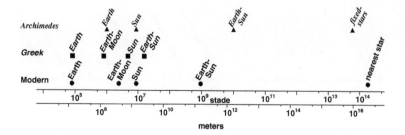

❧ 6 ❧

The Sand Reckoner

In the middle of the poem "The Walrus and the Carpenter," Lewis Carroll poses a nice little arithmetical problem. The Walrus and the Carpenter are walking along a seaside beach, gazing at the sand.

"If seven maids with seven mops
    Swept it for half a year.
Do you suppose," the Walrus said,
    "That they could get it clear?"
"I doubt it," said the Carpenter,
    And shed a bitter tear.

Through the Looking-Glass
Lewis Carroll

Who is right? Is it the optimistic Walrus who suspects that with a bit of backbone and elbow grease that the beach could be cleaned up and made into a tidy place? Or is the Carpenter right, that dower soul who totes a bag full of tools? Is the task too much? It is not very difficult to imagine that Carroll, in his other persona as a lecturer of mathematics at Christ Church college in Oxford, may have done the calculation himself. We, who have already calculated the number of cups in the ocean should not shy away from the question, "can the maids clean the beach?"

However, right from the first line I am struck with a problem. Can you really "sweep" with a mop? And if we assume we can, how much? If you were sweeping a floor you would expect less than a cup of sand in a minute. But if you swept a beach? I am going to be brazen here (we are talking about Lewis Carroll) and just postulate that they could sweep up between 1 and 10 l of sand a minute. Admittedly, the 10-l estimate might only happen if the maids traded in their mops for shovels. I do not think this too outrageous a suggestion, especially if the maids are enterprising and really trying to clean the beach in half a year.

Seven maids working for half a year could mean 8 hours a day, five days a week, which with a few holidays, adds up to about 1,000 hours per maid, or 7,000 maid-hours, or 420,000 maid-minutes. Alternatively it could mean seven maids continuously laboring away, or perhaps a platoon of maids working in shifts, twenty-four hours a day, seven days a week, for 182.5 days. That maximal labor is then 30,660 maid-hours or 1,839,600 maid-minutes. So at the low end of our estimate we expect that the maids could clear away 420,000 l or 420 m^3 of sand. At the other extreme (maids in shifts with shovel-like mops) they may remove 18,396,000 l or 18,396 m^3 of sand. This is a wide range of solutions, which we now need to compare to the beach itself.

Lewis Carroll gives us only a few hints as to the amount of sand there is on this beach. Later in the poem he tells us

The Walrus and the Carpenter
   Walked on a mile or so,
And then they rested on a rock
   Conveniently low:

So the beach is at least a mile long. Also, the presence of a rock may indicate that the sand is relatively shallow. If we use our maximal estimate of sand, 18,396 m^3, then it could be distributed as a strip 1,600 m long (1 mile), 1 m deep and 11 or 12 m wide. If, however, we use our minimal estimate of 420 m^3, that could cover a beach 1,600 m long, 10 cm deep and 2–3 m wide. To me this sounds like a pretty thin beach. Most beaches seem to me to be bigger than either of these estimates, and so I am suspect that the Carpenter is right. However, there are a few beaches where the Walrus's maids might just tidy things up a bit.

***

Lewis Carroll may be entertaining, but he really does not get us to large numbers. Eighteen million liters is not such an overwhelming number. Even if we measured it as three and a half billion teaspoons we can deal with these numbers. There are about a billion grains of sand in a liter, so this beach may contain a few quadrillion grains of sand, and even that number we can handle. But thinking about the number of grains of sand has often been associated with a vast or even uncountable number. For example, in the *Iliad*, Iris in the voice of Polites warns the Trojans about the Greek army, ". . . I have been in many a battle, but never yet saw such a host as is now advancing. They are crossing the plain to attack the city as thick as leaves or as the sands of the sea . . ." (Book II). Also, in the Bible, sand and stars are often seen as being uncountable, as in ". . . so many as the stars of the sky in multitude, and as the sand which is by the sea shore innumerable . . ." (Hebrews 11:12).

"There are some, King Gelon, who think that the number of sand is infinite in multitude." wrote Archimedes (287–212 BC) at the beginning of his essay *The Sand Reckoner*. I have always wondered what it was that prompted this treatise. I like to imagine that Archimedes and King Gelon, who apparently was interested in mathematics, were walking along a beach near Syracuse where they both lived. King Gelon may have said something as flippant as ". . . as countless or innumerable as the stars above or grains of sand on the beach . . ." and that this got Archimedes thinking "are they really countless?" I expect that Archimedes immediately recognized that the number of grains of sand was not infinite, but in some sense they were innumerable, because the number of grains of sand exceed a myriad-myriad, the largest number the Greeks could write in their number system.

Archimedes tells us in *The Sand Reckoner* that he has in fact devised a number system that could enumerate every grain of sand and he has already sent it off to Zeuxippus. But he also wanted to demonstrate this new system by calculating something really big, like the number of grains of sand on the beach.

There are some, King Gelon, who think that the number of the sand is infinite in multitude; and I mean by the sand not only that which exists about Syracuse and the rest of Sicily but also that which is found in every region whether inhabited or uninhabited. Again there are some who, without regarding it as infinite, yet think that no number has been named which is great enough to exceed its multitude.

So Archimedes sets himself a task, to calculate the amount of sand not only on all beaches, but if the whole universe was filled with sand. In fact he ups the challenge a bit; he did not want anyone in the future to say his number was too small, so at every stage he would overestimate the size of things to make sure his number was truly massive.

The reason we are going to spend a whole chapter with Archimedes is firstly he has an intriguing number system, and secondly because the way he estimates the size of the universe, even if it is really too small, parallels how modern astronomers work. But first, I will describe the Greek number system. When you learn a second language, it often leads you to better understand the structure of your first language; the same is true with number systems.

***

If you saw MDCLXXXVII you would immediately recognize it as a Roman number, and translate it into 1687, the year Isaac Newton published *Principia*. You can do that because you know the translation table (see Table 6.1).

The system we normally use is called the *Hindu–Arabic numeral system* or a *positional number system* with Arabic numerals. This system is a positionally based system, which the Roman system is not—or almost not. If you saw the first digit of *Principia*'s publication date, 'M', you would understand a value of 1000. In contrast, if you saw the first digit of 1687—the '1'—by itself you would think it meant "one." It is its position as the fourth digit as "1 _ _ _" that tells you that the meaning is "one thousand." The Greek number system is more like the Roman, but with a lot more symbols. In fact the Greek system needed twenty-seven symbols and so used all the characters of their alphabet, plus others.

In the Greek system, if I wanted to write out *Principia*'s publication date I would write ͵αχπζ′ (see Table 6.2). The little mark in front of

Table 6.1 The symbols and values of numerals in the Roman number system.

M	D	C	L	X	V	I
1000	500	100	50	10	5	1

Table 6.2 The symbols, values and names of numerals in the Greek number system.

α	1	alpha	ι	10	iota	ρ	100	rho	͵α	1000	alpha
β	2	beta	κ	20	kappa	σ	200	sigma	͵β	2000	beta
γ	3	gamma	λ	30	lambda	τ	300	tau	͵γ	3000	gamma
δ	4	delta	μ	40	mu	υ	400	upsilon	͵δ	4000	delta
ε	5	epsilon	ν	50	nu	φ	500	phi	͵ε	5000	epsilon
Ϝ	6	digamma	ξ	60	xi	χ	600	chi	͵Ϝ	6000	digamma
ζ	7	zeta	ο	70	omicron	ψ	700	psi	͵ζ	7000	zeta
η	8	eta	π	80	pi	ω	800	omega	͵η	8000	eta
θ	9	theta	ϟ	90	koppa	ϡ	900	sampi	͵θ	9000	theta

alpha is called a *low keraia* and is needed because we ran out of letters. α is 1, whereas ͵α is a thousand. The Greeks also needed a way to distinguish a word from a number and so they marked numbers with a ′ : a *keraia*. For example ροζ is a word meaning "pink" (try pronouncing rho-omicron-zeta, and you are speaking Greek), whereas ροζ′ is the number 177. Actually most of the time in classical Greece they would mark a number with an over-line instead of a keraia. There are also three Greek letters rarely seen, Ϝ (diagamma), ϟ (koppa) and ϡ (sampi). These letters had been dropped from use in words by the time of classical Greek, but were still needed to make the number system work.

With this system you can write any number from α′ (1) to ͵θϡϟθ′ (9999). The next larger number in this system has its own name, a *myriad*. A myriad specifically meant 10,000. Athens in the fifth century had about 40,000 citizens and probably a population of a few hundred thousand, and so a need to record larger numbers. For these purposes the Athenians would mark the digits above 9999 with a μ (mu) for myriad. So, for example in the number 12,345,678 they would first group as 1234,5678 and then write as:

$$\overline{{}_{,}ασλδ}\ {}_{,}εχοη$$
$$\mu$$

With this system they could count up to 99,999,999 or a *myriad-myriad*, which was as high as most Greeks needed. But that number was just the starting point for Archimedes. However, before we delve into how Archimedes proposed to extend this number system, we will look at his astronomy.

***

To measure the universe, either as a part of modern astronomy or cosmology, or to help Archimedes fill it with sand, we must start with the shape and size of the Earth. Once we know that, we can estimate the distance to the Moon, Sun and stars.

One of the myths we were taught as children was that Columbus had to fight the flat-earthers of his time and the fear that if you sailed into the far western seas you would eventually reach the edge of the Earth and fall off. Apparently the story of the fear of the edge of the Earth among Columbus's crew was primarily a construction of the storyteller Washington Irving. It makes a good tale but has little basis in fact: at the time of Columbus, most Europeans who thought about these things understood that the Earth was spherical.

Primitive societies, however, often started with a flat-Earth cosmology. When you look at the sea, large lakes, or great grassy plains or steppes they do look flat. Also, things fall off of globes but not table tops. A lot of people will tell you that they can see the curvature of the horizon while standing at the seaside, but that really is a tiny aberration, a barely measurable effect that is hardly noticeable unless you are certain it is there. A more dramatic effect is when you are sailing towards an island and you can see mountain tops while the shores of the island are still hidden below the horizon. Sailors speak of a ship being "hull-down," which means that it is so far away that you may be able to see the mast, but the hull is below the horizon. Still, one could argue that this is an optical illusion, like a mirage caused by heat or humidity.

Pythagoras (c. 572–c. 495 BC) may have been among the first to recognize that the Earth was a sphere, and within a few centuries of his discovery most Greek astronomers had come to the same conclusion. Aristotle tells us that the most convincing piece of evidence to support the idea is that the shadow of the Earth that falls on the Moon during a lunar eclipse is round. In addition, a lunar eclipse will appear later in the night to observers in the east then in the west. Since the eclipse is an event that is observed by everyone simultaneously, sunsets, sunrises and noon must happen earlier in the east then the west. Finally, the southern constellations drop below the horizon as you travel north. Most Greek astronomers found the evidence when put together overwhelming.

Once the shape of the Earth was established, the size of the Earth was a natural next question.

***

The most famous measurement of the Earth's circumference, and the one Archimedes cited, was made by Eratosthenes of Cyrene (c. 276– c. 195 BC). Eratosthenes was the third head librarian of the great library of Alexandria. Before its burning, this library was probably the greatest repository of knowledge in the ancient world. Contemporaries of Eratosthenes referred to him as "Beta" because he was not the top intellect in any one field, but he was near the top in many fields. His original writings have been lost to us, but we do know that they are based on him considering the city of Syene, which we now call Aswan, to be on the Tropic of Cancer. It is not clear exactly how he knew this, but Aratus, who described Eratosthenes' works, tells us that we will recognize the Tropic of Cancer because on the first day of summer, at noon, a gnome of a sundial, or any vertical column, will not have a shadow (see Figure 6.1). The Sun will be directly overhead in Syene (Aswan) while at the same time a gnome in Alexandria will have a shadow of just over 7°, or a fiftieth of a circle. Eratosthenes concluded that the distance between Syene and Alexandria was $\frac{1}{50}$ of the circumference of the Earth. The trek between the two cities had been measured at 5000 stade and so the circumference of the Earth, he concluded, must be 250,000 stade.

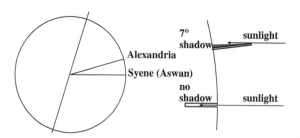

Figure 6.1 Eratosthenes' method for measuring the size of the Earth. This involved comparing shadows in Alexandria and Syene (Aswan), and knowing the distance between.

A book that has been lost for over a thousand years leaves a lot of questions. Which stade did Eratosthenes use? And how did he decide that Syene was on the Tropic of Cancer? Legends have grown up about Eratosthenes actually visiting Syene and looking down into a deep well on the first day of summer and seeing the reflection of the Sun. In fact on the island of Elephantine in the Nile river is a well now called Eratosthenes' well, a deep cool moist haven in the hot dry Egyptian desert. But we know of no ancient connection between Eratosthenes and this well.

A lot of people have argued about what that 5,000 stade from Alexandria to Syene means. As was mentioned in Chapter 1, there were many different stades in use in the ancient world. Was it the Olympic stade (176 m), the Roman stade (185 m), the itinerary stade (157 m) or some other unit? The 5000 stade number itself is probably pretty good. At the time there were professional "pacers" who measured long distances. But he may have also just asked the opinion of caravans who had trekked the distance. Most people who study this problem use the itinerary stade to convert Eratosthenes' 250,000 stade into 39,000 km, which is just a bit less than modern measurements.

By modern measurements, Alexandria is at latitude $31°12'$ N and Aswan is at $24°5'$ N, so the difference is just over $7°$, and a bit more than a fiftieth of a circle. The distance about 840 km, although not all of that is north–south. Dividing the 840 km by the 5000 stade Eratosthenes reported would suggest a stade of 168 meters, something in the middle of those cited above. I actually think the number 5000 tells us that Eratosthenes understood that this was not an exact measurement. It was rounded off and only a rough attempt to describe the size of the Earth.

Eratosthenes' measurement is within about 10% of our modern one, but more important for this chapter, it forms the basis for measuring Archimedes' cosmos. However, Archimedes wanted a large universe, so he rounded 250,000 up to 300,000 and then multiplied by 10 to make sure he did not underestimate.

***

The next step is to measure the distance to the Moon, and the Greeks did a good job on this too. However, measuring the distance to the Moon is a bit different than the distance around the Earth. You cannot pace off even part of the way.

When I was young I remembered seeing a chart in my classroom that showed the heights of different types of clouds. Cumulus were at less than 2000 m whereas cirrus were at a much dizzier height of 7000 m. I also learned that people knew this before airplanes or even hot air balloons and was left with the question of how they had measured these distances. Years later, on a beautiful summer day, I was sitting on a hilltop watching clouds when the secret of how to measure their heights came to me. First, I could figure out how big the clouds were by looking at their shadows: that big puffy one that reminds me of the face of a moose cast a shadow on that far hill, which goes from that big oak tree to the Holstein grazing in the western pasture. If that cow would remain still I could run over to the hill and pace off the distance between the tree and Holstein, and so measure the size of the cloud. Once I know its size, by measuring a few angles and using trigonometry I would be able to calculate its altitude, thickness and any other dimension. An earthbound observer could in fact measure the height of this non-terrestrial body.

In a similar way Aristarchus of Somos (c. 310–c. 230 BC) measured the size and then the distance to the Moon. He started out by watching a lunar eclipse and noted at what time the Moon started to enter the shadow of the Earth, at what time it was completely in the shadow, and at what time it emerged from the other side (see Figure 6.2). He reasoned that since the Sun was very far away, the distance across the shadow of the Earth is about the diameter of the Earth. Then the ratio of the time to enter the shadow to the time to cross the shadow was about equal to the ratio of the diameter of the Moon to the diameter of the Earth. From that he deduced that the diameter of the Moon was about a third that of the Earth.

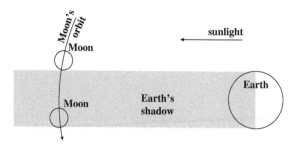

Figure 6.2 Aristarchus's method for measuring the size of the Moon.

Curiously enough, when Aristarchus proposed this technique he could not report the distance to the Moon in stade because Aristarchus preceded Eratosthenes and so he did not know the size of the Earth. Of course Archimedes followed both of them and so had access to both Aristarchus's technique and Eratosthenes' baseline: the diameter of the Earth.

Just now I introduced the term "baseline." We are going to see a lot of baselines in this and later chapters as we survey space. So it is worth taking time now to try and understand what a baseline is. Some of the techniques used to map space are the same as the techniques used to map the surface of the Earth, especially when trying to measure the distance to a place you cannot touch. Let us imagine that you would like to measure the distance across the Grand Canyon. You are not going to stretch a tape measure from rim to rim. Also, hiking down the Bright Angel trail on one side and up the north Kaibab trail on the other side, with its switchbacks and serpentine route, is not going to tell you the straight-line distance. However, you can survey the distance, using geometry and trigonometry.

You can stand on Yaki point, near the park's visitor center on the south rim and note that due north, on the other rim, is Widforss point (see Figure 6.3). Now you can hike west on the relatively straight and level trail for about 5.5 km to the Bright Angel trail's south-rim trailhead. From there you can again sight Widforss point and see that it is now north-by-northeast of you, or 23° from north. So, if you sketch this out you have a triangle with Yaki point, the Bright Angel trailhead and Widforss point. Since you know two angles, 90° at Yaki Point and 67° at the trailhead, the last angle at Widforss must be 23°. You know everything about the shape of the triangle, but you do not know the size until you remembered the 5.5 km you just hiked. That one distance sets the scale of the whole triangle. This one distance that you measured directly is the baseline of the survey, and the distance across the Grand Canyon, or at least from Yaki to Widforss Point, can be determined at about 13 km.

The baseline to measure the Earth was the distance from Alexandria to Syene. The baseline to measure the distance to the Moon was the diameter of the Earth. The baseline to measure the distance to the Sun is the distance to the Moon. All measurements of great distance build upon previous, smaller measurements, something which is true even today when we measure the distances to galaxies.

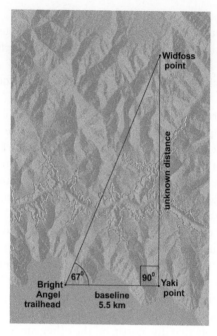

Figure 6.3 Surveying the distance to unreachable points with angles and baselines.

The measurement of the distance to the Sun is hard, but again it was Aristarchus who provided us with a method. His technique was to look at the Moon and note the moment at which exactly half the Moon is illuminated (see Figure 6.4). He reasoned that this was the moment that the Earth–Moon line was at right angles to the Moon–Sun line, so

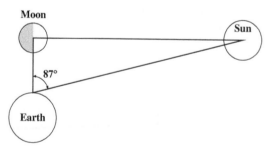

Figure 6.4 Aristarchus's method for measuring the size of and distance to the Sun.

the Earth–Moon–Sun angle was 90°. At that same moment he could measure the Moon–Earth–Sun angle as 87°. So he had a triangle, he knew two angles and a baseline, and so he could calculate the distance to the Sun.

Mathematically, this was a beautiful solution. Experimentally, it was very hard to carry out. First and foremost it is difficult to determine exactly when the Moon is half full. Aristarchus picked his moment, measured his angle, and found that the distance to the Sun was 190 times the diameter of the Earth. But he was a few hours too early. Ideally he would have measured 89.5° but as the angle approaches 90° a few minutes makes a huge difference. At 90° the mathematical distance is infinite.

In fairness to Aristarchus I have oversimplified his method. He recognized that because the Sun was not infinitely far away, the shadow of the Earth would taper and be smaller than the Earth when it reached the Moon's orbit. To correct for this he combined the two measurements, and added one more observable, the angle size of the Moon and Sun. Aristarchus underestimated the size of the Sun by a factor of six, and the distance to the Sun by a factor of about fifteen.

Archimedes was well versed in this method and followed it closely, but then offered one original measurement. He devised an instrument made of cylinders and a measuring rod with which he determined the angular size of the Sun and Moon. He found that they were a thousandth of a great circle, or about 0.36°, close to the 0.5° we now measure. This information would have helped Aristarchus's calculation. However, Archimedes was already trying to round things up and get the greatest distances he could, ending up with a distance to the Sun of 100 myriad-myriad stade (10^{10} stradia), or about ten times our modern measurements.

For Archimedes there was one more step. How big is the cosmos, or how far away are the stars? In the end we do not really know how far away Archimedes thought the stars were; we just know he was looking for a huge dimension, and to do that he did something curious. He had several cosmological models available to him: the geocentric model where everything orbits the Earth, the heliocentric model where everything orbits the Sun, and the semi-heliocentric model where Mercury and Venus orbit the Sun and the Sun orbits the Earth. He picked the heliocentric model of Aristarchus, not because it was popular (most astronomers at that time did not like it) or because he personally believed in it (he does not tell us what he thought), but rather because it

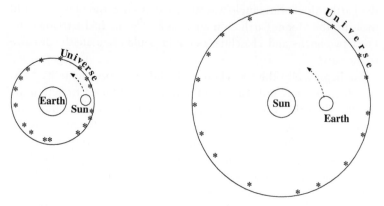

Figure 6.5 The size of a geocentric and a heliocentric universe. (Left) In a geocentric universe the stars do not need to be much beyond the Sun. (Right) In a heliocentric universe the stars must be far beyond the Sun, since we do not see them shift as the Earth moves.

gave him the largest cosmos. In a heliocentric cosmology the Earth is in motion around the Sun, tracing an orbit of tens of millions of stade in diameter, yet the stars are so distant that we do not see them shift (see Figure 6.5). They must truly be very far away.

What Archimedes was going to do next was to look for parallax. The idea of parallax is that nearby objects seem to shift compared to a background when the observer moves. For instance, hold up a finger at arm's length and view it with one eye, then note its position compared to a distant wall or landscape. Now close that one eye and open the other. Your finger may not have moved, but it now shows up against a different part of the background. The effect just described is parallax, a type of surveying where the distance between your two eyes forms the baseline of the measurement. When measuring with parallax the longer the baseline, the greater the shift.

What Archimedes does with parallax is curious. First he notes that no astronomer has observed solar parallax. This would be the shifting of the Sun in the sky due to the observer's motions. Every day our position changes by a distance equal to the diameter of the Earth as the Earth rotates and moves the observer from dawn to dusk. Also, no stellar parallax is observed as we move around the Sun over a six-month period. Therefore, Archimedes concludes that the ratio of the

Sun-to-stars to the Earth-to-Sun distances is the same as the ratio of the Earth-to-Sun distance to the Earth-diameter. This gave him a magnificent universe with a radius of about 10^{14} stade or 100 quadrillion m! That is an amazing leap of the imagination. The distance also happens to be close to about 1 lightyear, or a third of the distance to the nearest star by modern measurements (see Table 6.3).

***

The distance to the fixed stars was not what Archimedes was looking for. He really wanted a big number and would have liked to count the number of grains of sand that would fill all of space. He said that 10,000 grains of sand—one myriad—was equal to a poppy seed and that 40 poppy seeds side by side was a finger's breadth. From this we know that the sand he was referring to was about 2×10^{-5} m ($20 \mu m$) across. This is the type of sand used in 500 grade, "superfine" finishing sandpaper. In fact you will rarely find sandpaper this fine. It is more like the grit on emery cloth. Still, if you were to place 7×10^{20} of these superfine grains of sand side by side they would reach Archimedes' fixed stars. Or, if we were to fill the whole cosmological sphere with sand we would need 10^{63} grains. That is the large number Archimedes was striving for.

***

The normal Greek number system can only count up to a myriad-myriad. What Archimedes did to deal with this limitation was to classify all numbers up to this point as the "first order." Bigger numbers were of higher orders. So, for example, the population of the Earth today is about seven billion, or

$7,000,000,000 = 70,0000,0000$
$= 70 \times 10^8 = 70 \times$ myriad-myriad
$= 70$ of the second order.

Or consider Avogadro's number

$6.022 \times 10^{23} = 602,200,000,000,000,000,000,000$
$= 6022,0000,0000,0000,0000,0000 = 6022,0000\ 0000,0000\ 0000,0000$
$=$ six-thousand twenty-two myriad of the third order

This system could easily "name a number" up to a myriad-myriad orders, or in modern notation, $10,000,000^{10,000,000}$.

Table 6.3 A comparison of the astronomical measurements of Aristarchus, of Archimedes and of modern astronomical measurements.

Object	Greek (stade)	Archimedes (stade)	Greek (km)	Archimedes (km)	Modern (km)
Earth (circumference)	250,000	3,000,000	40,000	480,000	$40,074.5^p$ $39,940.6^e$
Earth (diameter)	80,000	1,000,000 $100\,\mu$	13,000	160,000	$12,756.1^p$ $12,713.5^e$
Moon (diameter)	$\frac{1}{3}D_\oplus$ 25,000	$\frac{1}{3}D_\oplus$	4,000		3,476
Earth to Moon	$10D_\oplus$ 800,000	$10D_\oplus$	128,000		384,403.1
Sun (diameter)	$19D_\oplus$ 4,750,000	$30D_\oplus$ 90,000,000	760,000	1,440,000	1,391,000
Sun (angle)	$\frac{1}{15}$ of zodiac $(2°)$	$\frac{1}{1000}$ of circle $(0.36°)$			$(0.5°)$
Earth to Sun	$190D_\oplus$ 10^7	$4775D_\oplus \sim \mu D_\oplus$ $\sim 100\,\mu\mu \sim 10^{10}$	2.5×10^6	1.6×10^9	149,669,180
Sun to stars		$22,797,562D_\oplus$ $\sim 10^{14}$		1.6×10^{13} ~ 1ly	1.5×10^8

"Greek" refers to measurements of Aristarchus of Somos. p, polar; e, equator; μ, myriad (10,000). I use 1 stade = 160 m = 0.16 km.

However, Archimedes did not stop there. All the numbers up to a myriad-myriad order are called the first period, which of course means that there is a second period, a third period and so on up to a myriad-myriad period. We could write that number as

$$10,000,000^{10,000,000^{10,000,000}}$$

a number for which even Archimedes did not have a use.

There is actually a parallel between Archimedes' system and the number system that we presently use, especially when we think about the naming of numbers. Earlier on in this chapter I said that learning a second language or number system helps us understand our first one better. If we were to write out Avogadro's number as words we would write "six hundred two sextillion, two hundred quintillion." We could read sextillion and quintillion as "of the sixth -illion" and "of the fifth -illion." This actually makes more sense in the long-scale number system where million, billion, trillion and quadrillion are powers of a million. That is, in the long scale, billion = million^2, trillion = million^3 and so forth. That means that "-illion" mimics Archimedes' "order."

This observation actually brings us to the limit of Archimedes' system. His language would not support anything larger. When we write a trillion, we mean the third -illon. Here 'third' is an adjective that we have made out of 'three', which is a noun. So the largest normal name for a number is *vigintillion*, which is 10^{63}; coincidentally Archimedes' sand count. *Viginti* is Latin for twenty, so vigintillion means $3 \times (1 + 20)$. Beyond twenty the names of numbers are not unique. However, in English we can take any number and make it into an adjective, something Archimedes could not do. Still, his number system, even just the first period, is not only sufficiently large to count all the sand that could be crammed into a sphere with a radius of 1 lightyear, it could also count all the quarks in the modern observable universe, or all the photons. In fact, if we filled the observable universe with particles each a Planck length in diameter, Archimedes could still name that number.

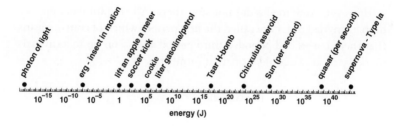

Energy

Energy is one of the most common and confusing topics we will consider in this book. It is ubiquitous, touching upon almost every aspect of our lives. We see it in transportation and food, in warmth and lighting and a hundred other areas. But precisely because it is omnipresent we almost fail to see it. Or we do see it but as a hundred different things. It is the forest that is lost in the trees. There are so many ways to describe energy, so many ways to meter it out. A cord of wood contains 15–30 million BTUs of energy. A cookie might contain a few dozen *Calories*, while the energy needed to boil water for a cup of tea to go with that cookie is about 2000 *calories*. You may have noticed in the last sentence that the food Calories are not written the same way as the heat calories. You pay an electric bill in kilowatt hours. All of these various units are everyday and human scale.

The destructive energy of a large bomb is measured in *megatons*. The energy released when a neutron decays into a proton is just under an MeV. Astronomers report that a quasar can spew forth about 10^{45} *ergs* each second. Each of these quantities is reported in their own units, each is measured in its own way. But those units, the calories or BTUs or ergs, actually tell us something about how that energy is measured or the history of that type of fuel or the expected outcome of releasing that much energy. But we can, and will, sort it all out and find the Rosetta stone of energy. This sorting and cataloging of apparently unrelated facts is what science does well when it is working at its best. Taking a list of chemical elements and their properties, Dmitri Mendeleev formed them into his *periodic table* of the elements. Taking a list of plants and their anatomy, Carl Linnaeus organized them into his *binomial system*. But the true genius and value of these systems is that they reflected a deeper aspect of nature. The rows and columns in Mendeleev's table foretold the structure of the atom and Linnaeus's system reflected evolutionary history.

So we will take Linnaeus and Mendeleev as our inspiration and expect that out of apparent chaos, order will arise. We will take our energy factoids and stir them about. We will walk about them like a sculptor eyeing a block of marble, trying to see the statue inside. We will rearrange them until a pattern emerges at which point we expect to see something deeper and more universally significant.

In the end I would like to plot a diagram, as I have done in most other chapters, which has objects with low energy at one end and high energy at the other end. And in the process we will create order and learn something.

So what is energy? It is one of those things that at first is hard to define but, like beauty, you know it when you see it. So let us start with a event that we will all agree contains a lot of energy.

***

On the front page of the sports section of a newspaper is a color photograph from a soccer match. In the middle of the picture is a soccer ball, frozen in time such that we can see the black pentagons and white hexagons that make up the ball. Also frozen in the air, with his fingers 10 cm from the ball, is the goalkeeper. He is wearing a yellow jersey and is apparently diving towards the ball with outstretched, gloved fingers. Finally, in the foreground, is a striker clad in sky blue. He too is frozen in time, with one foot stretched out in front of him, toe pointed towards the ball. Clearly there is a lot of energy in this scene. Or is there?

What if the whole picture had been staged?

What if the ball and the keeper were suspended from thin wires? What if the striker had ballet-like balance and could stand on the toe of one foot? In that case the scene might in fact be *static*, with no motion and so no energy. The photograph is crystal clear. Under a magnifying glass we can see the pixels from the printing press, but we cannot see any blurring to tell us about motion.

Actually there is one more piece of information available to us on the page. We know that the event is important, even newsworthy, since it is on the front page of the sports section. Not only was there a shot on goal, but it must have been a pivotal moment in the match. So, did he score or did the keeper defend his net?

If we also had a photograph from a tenth of a second later all would be clear. The keeper crashing into the turf, the striker with a triumphant glow, the ball 3 m forward and well within the net. There indeed was a lot of energy in that photograph.

Three meters in a tenth of a second tells us that the ball is rocketing along at 30 m/s (over 100 km/h or 60 mph). Since the ball must meet FIFA standards it must have a mass of 410–450 g, which also means it has an energy of about 200 J. Finally, we have met the most basic unit of energy: the joule (J).

When we talk about motion, like a soccer ball in flight, energy is clearly involved. However, it might be useful to start out with a motor or engine to get a handle on energy; engines tend to have steadier and more regular paths than soccer balls, and when we talk about engines they tend to be rated in terms of watts or, more traditionally, in terms of *horsepower*. Now we are back again to all these charmingly and curiously named measurement units, with their own stories. Horsepower was a term that was coined by James Watt (1736–1819) a Scottish engineer who vastly improved the steam engine. Watt was trying to sell his engines to mines to pump out water. Prior to the steam engine, pumping had been done by horses walking on a mill wheel (like a big gerbil wheel). Watt determined that one horse could produce "33,000 foot-pounds per minute," which is the same as a horse raising 33,000 pounds one foot in one minute, or 55 pounds a distance of foot in a second, or 11 pounds up 5 feet in a second, or many other combinations.

But that is not energy.

James Watt also lent his name to another metric unit, the *watt*. That too is not energy. The horsepower (hp) and the watt (W) are units that measure *power*. A horse will produce one horsepower. If it works a minute or a day, it is one horsepower. But clearly if it toils all day it is using a lot more energy than if it only labored a single minute. Energy is power exerted, multiplied by the amount of time involved. In terms of an electric bill, a *kilowatt hour* is the power of a kilowatt, used for one hour. Conversely, power is the rate at which energy is used.

So that soccer ball with 200 J of energy has the same energy that a 40 W light bulb uses in 5 s. Or, since the striker got the ball moving in a tenth of a second, during that brief contact he exerted about 2000 W.

James Prescott Joule (1818–1889), the person for whom the energy unit is named, was a brewer and an active scientist who lived in Manchester, England. He was interested in the relationship between heat and energy and in the middle of the nineteenth century he quantified their relationship. He performed a series of experiments that involved putting a measured amount of energy into water and seeing how the temperature of the water changed. These included using electrical energy and compression energy. But his most famous experiment involved the vigorous stirring of water—the use of kinetic or mechanical energy—to get its temperature to rise. To do this he placed a small paddle wheel in the water and attached the wheel to a string. The other end of the string had a weight that he could release, thus producing energy in a very controlled way. The amount of heat he produced was minuscule, a fact that left some of his contemporaries skeptical. In fact, recent reconstructions of his experiment have suggested that the only reason that he could measure this temperature change was because of his experience with brewing. Perhaps the most important thing about Joule's experiments was that three very different experiments got nearly (to about 10%) the same conversion factor between energy and heat. In modern, metric units, one calorie of heat is equal to 4.184 J of energy.

These results, plus those of a number of Joule's contemporaries, finally pushed us toward the principle of the conservation of energy: energy is never created nor destroyed, but it can change its form. As James Joule wrote:

Believing that the power to destroy belongs to the Creator alone, I entirely coincide with Roget and Faraday in the opinion that any theory which, when carried out, demands the annihilation of force, is necessarily erroneous.

James Joule, 1845

In this case *force* means heat or energy.

Once it was established that energy was conserved, it was possible to talk about the flow of energy. We can see that the energy of a steam engine can turn a generator that produces electricity that can pass through a resistive wire and get hot, perhaps heating more water and making more steam. We know that it is not perfect. At every stage some of that energy is transformed into entropy, a form of energy that is not useable to us. But we should also ask what it was at the start of the flow that heated the water. Where did the original energy come from? We need the initial heat from a burning lump of coal or chunk of wood. But the piece of raw wood is not hot or in motion to start with. Where is its energy?

***

What we have in a piece of wood, or a bucket of kerosene is *potential energy*; it is potential as in "unrealized" or "possibly in the future;" it *may* cause motion or heat. But it really is energy right now too, even if it is not in motion or heat. Potential energy includes things like the weights on a cuckoo clock or on Joule's water-stirring machine. Energy is put into the clock at the beginning of the week when the clock's caretaker raises the weights. That energy is released a bit at a time when the weights slowly fall and the hands on the face of the clock spin.

There is also energy in a sandwich. This can give us the energy to climb a hill or wind a clock. There is potential in springs and batteries. When Sisyphus, from Greek mythology, rolled his stone up a hill in the underworld, he was putting potential energy into it, energy that became motion when the stone rolled back down. All food and all fuel contains potential energy, including hydrogen, the fuel our Sun burns, a uranium nucleus that might decay, and an asteroid, which may plummet to the Earth.

***

Let us step back a moment and try to see where we are headed. We are looking for an organizing principle for energy, something that will take oil and speeding comets and let us plot them relative to each other and see where energy is fundamentally big and where it is small. We now have one organizing strategy. We can categorize energy as motion (kinetic), heat, or potential. But that does not effect the plot of big energy and little energy in a fundamental way. We can sort into three groups, but we do not have a real basic scheme. So we need to keep hunting and looking at energy and trying to find a pattern.

***

Lets go back to the idea that energy flows and see where it carries us. The source of nearly all energy in our solar system is the Sun. Within the

Sun, two protons in hydrogen atoms collide and form a helium atom, which then settles into a deuterium atom (an isotope of hydrogen) and in the process produces about 2×10^{-13} J of energy. That is not a lot of energy, but this sort of reaction happens in the Sun about 10^{39} times each second.

As a whole, the Sun radiates about 4×10^{26} W or 4×10^{26} joules of energy per second. This energy is carried outward by the motion of particles that have been boiled off the surface of the Sun, which we call the solar wind. It is also carried off as radiation: light, radio waves, X-rays, ultraviolet, infrared and other forms. All this energy streams outward in every direction.

From the Sun's perspective, the Earth occupies only a very tiny spot in the sky, so the Earth only intercepts a very tiny fraction of that outflow of energy, about 1.76×10^{17} J/s. By the time the light has come through the atmosphere about 30% has already been reflected back into space, in particular, most of the ultraviolet. Still, about 1000 J/s of energy strike each square meter of the Earth, if it is a cloudless day and the Sun is directly overhead. Modern commercial solar cells are about 15% efficient or perhaps a bit more, with laboratory cells measuring nearer to 40%. So if our square meter was covered with an average panel of solar cells we could expect about 150 J of energy every second, or 150 W of power. With that we could charge up batteries for our lights at night or drive a solar race car. So we have gone from the potential energy locked in the fuel of our star to the chemical potential in a battery. But that same sunlight can do a lot of other things as well.

Most of the Sun's light ends up heating the Earth. Because the Earth has a steady temperature (climate change is slow compared to the amount of energy that heats the Earth daily), all this energy eventually is radiated away. But for a while it dwells on Earth, drives the winds and stirs the oceans by heating different parts of our planet differently. Obviously the equatorial regions, with the Sun poised directly overhead, catch the most energy and are well heated, whereas the polar regions, with slanting sunlight, are only slightly warmed. In fact about 1% of the energy from the Sun ends up driving the wind and the currents on Earth.

The sunlight can also smile upon plants. A little less than 0.08% of the energy in sunlight is absorbed by plants via the process of photosynthesis, transforming that energy into chemical energy embedded in the plant itself. We think of biofuels as the energy in the oils or starches of a plant, but the whole plant—the stalks, the roots and the leaves—is potential energy. The Earth produces about 150 billion tons of biomass each year. And all of that is potential energy.

Over 20% of the Sun's heat also drives the water cycle; evaporating water mainly from the oceans and then raining everywhere. If the rain falls on highlands we can collect it and run it through turbines that spin generators and produce electricity. If we are not tapping that energy, nature itself is using the rivers to sculpt mountains and valleys and to build up flood plains and deltas.

At first it appears as if all of our energy comes from the Sun: heat, biomass, wind and hydro. Even the energy in fossil fuels, which is no more than ancient biomass, comes from the Sun. However there are a few non-solar energies in play on Earth.

Nuclear power stations may be the most obvious non-solar energy form on Earth. The uranium that fuels these plants predates the Sun. The rarer elements resulted from a supernova, the death convulsions of an ancient star before our solar system even formed. But human-built power plants are not the only way to tap into this power source. The fact that the center of the Earth is at about 5400°C is in part due to radioactive material deep within our planet. It is not that there is a reactor with neutron-induced fission going on but rather a few atoms, sparsely distributed, are undergoing spontaneous radioactive decay and in that process giving up heat. It is not a lot of heat per decay. In fact there is no more heat in a kilogram of rock deep in the Earth than one on the surface. But there is a huge amount of material underfoot, and all that heat eventually must come up through the surface. It is like the city described in Chapter 2. There the volume of the city grew faster than its surface area, and so the traffic on the arterial roads became more intense. We see that energy traffic, that outflow, as volcanos, geysers and hot springs.

One final non-solar energy source here on Earth is tidal power. It has only been tapped by humans in a very few unique locations, such as in the Bay of Fundy in Canada and Rance River, near St. Malo in France. Yet if you have ever watched the ocean rise and fall and contemplated the trillions of tons of sea water that is being shifted, it is clear that there is a lot of energy here.

***

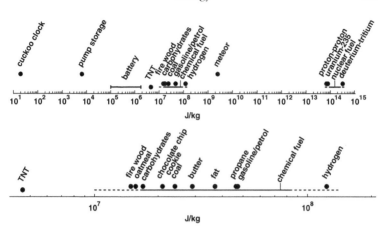

Figure 7.1 Energy densities for chemical and nuclear fuels. (Top) Energy densities. Chemistry and nuclear fuels are well separated. Gravity does not really fit and so here it is offset. (Bottom) Detail of chemical fuel densities.

Flow was an interesting way of looking at energy, and in the process we touched on a number of examples, but we still do not have that organizing principle. However, I think I now know what I do not want to plot. I know that a piece of wood contains a million joules of energy whereas a cord of firewood contains a billion joules. I do not want that to appear twice on our plot. This is like in Chapter 2, where I did not plot all elephants or whales, but rather only the largest. I would like to plot the energy for wood and the energy for uranium and the energy for a comet, and look for a single number that characterizes all pieces of wood or uranium or celestial ice balls, independent of their varied sizes. To do this I will concentrate on energy density (see Figure 7.1). How much energy does a gram of fuel or a flying soccer ball contain.

***

According to the nutritional panel on the side of my oatmeal box, one serving starts with 40 g (half a cup) of dry oatmeal and contains 150 Calories. These are food calories, and we will now return to the maze of energy units. One food calorie is one thousand heat calories, or one kilocalorie; the upper case "C" is just shorthand for this. So my 150 breakfast Calories is 150,000 heat calories or about 630,000 J, something we know because of James Joule's experiments. So we find that oatmeal has 16,000 J/g.

Pure carbohydrates and proteins are a bit higher that raw oatmeal, with 17,000 J/g, but fats and oils pack about twice the energy density, with about 37,000 J/g. A chocolate chip cookie, which is a combination of carbohydrates, proteins, fats and oils is someplace in the middle, at about 20,000 J/g, depending upon the particular recipe. Curiously enough, TNT only has about 3,000 J/g. It is hard to believe that a cookie actually packs seven times the energy of TNT, yet we do not use cookies to blast rocks, because of how fast they burn.

Non-food fuels run across a wide range. But here again we have to pause. If you look up the energy in wood you might see it reported as "24 million BTUs per cord" for red oak. A cord is really a volume, defined as a stack of firewood 4 feet wide, 8 feet long and 4 feet high. The cord is a unit that makes sense to people hauling wagonloads of firewood. It can be converted to pounds and kilograms. But what about BTUs?

A BTU is a *British thermal unit*, and is defined as the amount of heat that will raise 1 pound of water by 1 degree Fahrenheit. Its definition is very similar to a calorie, which is defined as the amount of heat that will raise 1 g of water by 1°C. Since there are 1000 g in a kilogram this also implies that 1 Calorie is the amount of heat to raise 1 kg by 1°C.

Finally we can convert that "24 million BTUs per cord" for red oak into an energy density and find a figure of just under 15,000 J/g, very close to the value for oatmeal. People who burn wood tend to prefer oak, maple and ash over pine and hemlock because there are more BTUs per cord (and also because pine can be sooty). But a cord of oak actually weighs more than a cord of pine, and the energy density across all types of wood is nearly constant at 15,000 J/g.

The energy density of coal is a bit higher than wood, at 24,000 J/g. Petroleum-based fuels all weigh in at about three times that density, with propane at 46,000 J/g and gasoline-petrol at 47,000 J/g, which is to be compared to the density of food fats and oils. Still, all of these fuels vary by only a factor of three.

Nuclear is on a scale of its own. Uranium-235 packs 79,500,000,000 J/g (7.95×10^{10} J/g), an energy density about 2 million times greater than gasoline. That is also very close to the density of energy in hydrogen if it is undergoing fusion, as in the Sun. An ideal fuel would be deuterium and tritium, with an energy density of 3×10^{11} J/g.

What about the weights on a cuckoo clock? At the beginning of the week, when the weights are raised, these may have an energy of 20 J/kg or 0.02 J/g. That might be enough energy to spin the hands of the clock, drive the bird out to recite its cuckoos and chime the hours, but it is not a lot of energy.

Another way in which gravity is used to store energy is with *pump storage* facilities. In these places water is pumped uphill to a high-elevation reservoir to store energy. When the energy is needed the water is released and it runs downhill, through a hydroelectric turbine that converts that gravity potential into electrical energy. Some of the largest of these are Bath County facility in Virginia and the Kannagawa project in Japan. Both of these can produce about 3 million watts of power, for about ten hours. Ten hours is a useful amount; it means you can store excess energy at night, when other power plants still produce it but there are few customers to consume it, and then use it in the day when the demand is high. As far as energy density is concerned, the Kannagawa project is the more impressive. It uses less water, but pumps it higher and achieves an energy density of 6.4 J/g.

When we plot different systems or objects or fuels as a function of their energy density we find that on the far left are things driven by gravity, in the middle are things with energy stored in chemical bonds, and on the right are nuclear fuels. Each group is separated by a factor of nearly a million. It seems like a trend that will help us organize energy and also fits with our instinctive feeling that gravity is weaker than chemical bonds and that nuclear forces are stronger than everything else. But how do we reconcile this with the fact that a falling meteor contains up to 2.5 million J/g?

***

The meteor that struck the Earth about sixty-five million years ago had about 4×10^{23} J of energy. This was the meteor that formed the 180-km-wide Chicxulub crater, on the Yucatan peninsula in Mexico. Its impact and the resulting dust are credited with causing the demise of the dinosaurs. However, it is unfair to say that it was a typical meteor; it was a chunk of rock about ten kilometers on a side. Still, there are some things common to all meteors.

If I were to drop a ball from the edge of the universe, by the time it hit the Earth it would be traveling at about 11 km/s. It does not matter the size of the object I drop, much like the two iron balls Galileo dropped from the top of a tower. If I drop a golf ball: 11 km/sec. If I drop a bulldozer: the same. But that is not up to the speed of meteors.

If I dropped these same objects on the Sun they would be traveling at 62 km/s when they hit the Sun's surface. This is because the Sun is much more massive than the Earth and exerts a much greater gravitational force. This is why meteors that hit the Earth average about 42 km/s, which seems very different to the 11 km/s cited in the last paragraph. Actually, meteors that hit the Earth are generally being attracted towards *the Sun*, but the Earth gets in their way. Their velocity can also be modified by the fact that the Earth is in motion around the Sun at about 29 km/s, and part of that velocity could add to or subtract from the 42 km/s of the meteor depending upon the relative directions of their motions.

So the energy related to gravity depends upon the masses of the attractor and the attractee. The binding of the Moon or a satellite to Earth does not really characterize gravity in a general way; rather, it characterizes the particular nature of gravity on Earth.

It is true that a rock falling to Earth releases potential energy just as the burning of wood or oil does, but it is not really a fuel. A fuel is a material, a substance, that contains energy intrinsically in its structure. Fuels also are stable. They just sit there until they are somehow ignited.

When we burn a piece of wood or a teaspoon of gasoline, or when we metabolize a bit of bread, we are releasing energy by breaking chemical bonds. As an example we will look at the burning of hydrogen and oxygen;

$$2H_2 + O_2 \longrightarrow 2H_2O + \text{Energy}$$

We start out with three molecules: two molecules of hydrogen (H_2), each with two atoms (remember Dalton and Avagadro?) and a oxygen molecule (O_2), as shown in Figure 7.2. These molecules are stable and will just sit there. I have to apply some energy to break them up. It takes 4.5 eV of energy (yes a new unit; just think energy on the atomic scale) to break up $H_2 \rightarrow H + H$ and 5.1 eV for $O_2 \rightarrow O + O$, or 14.1 eV to break up all three molecules. Once they are broken up they will likely recombine into $2H_2O$ and release 20 eV in the process. So there is a net release of about 6 eV. I went through this example to show why fuels do not burn spontaneously. They need a spark to get the process going.

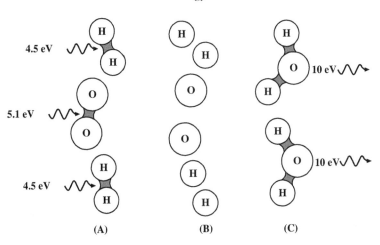

Figure 7.2 The energy that is released when hydrogen burns, at the molecular level. (A) Hydrogen and oxygen are broken by external energy. (B) Atoms regroup. (C) When the atoms recombine into H₂O molecules, extra energy is released.

If we turn to more complex fuels, like wood or oil, the molecules involved are different, but the principle is the same. Most foods and fuels are hydrocarbons, meaning that the molecules involved are primarily made of hydrogen and carbon. Some fuels, such as glucose ($C_6H_{12}O_6$), also have some oxygen. One of the most important and well-known molecules in petroleum is octane. Octane chemically is C_8H_{18}, often written as $CH_3(CH_2)_6CH_3$, which is a chain of carbon atoms, with hydrogen filling any unfilled carbon bonds (see Figure 7.3). When octane burns it ideally takes two octane molecules and twenty-five oxygen molecules for the full reaction.

$$2\,C_8H_{18} + 25\,O_2 \longrightarrow 16\,CO_2 + 18\,H_2O + \text{Energy}$$

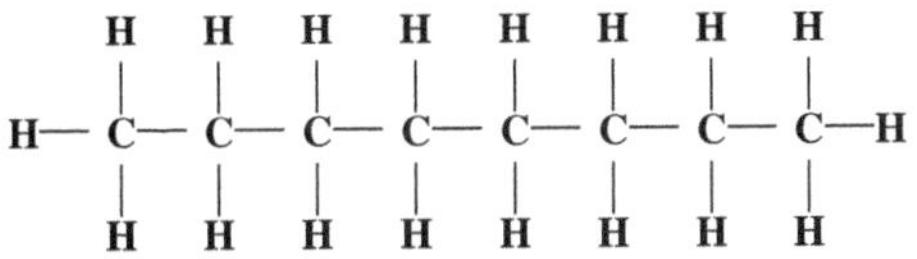

Figure 7.3 The octane molecule. Octane is a hydrocarbon chain made up of eight (oct) carbon atoms and eighteen hydrogen atoms, which means that there are a lot of bonds with energy stored in them.

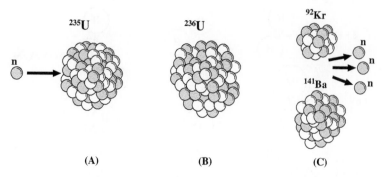

Figure 7.4 Energy from nuclear fission. (A) A neutron hits ^{235}U. (B) The uranium is excited into ^{236}U, which is unstable. (C) ^{236}U decays into ^{92}Kr + ^{141}Ba + 3n + Energy.

What makes octane such an attractive fuel is that it has a lot of chemical bonds and energy in a compact form.

Nuclear fuel is similar. For example, the most common fuel in a nuclear power plant is uranium-235 $\left(^{235}U\right)$. It, like octane, will just sit there for a long time. However, in a reactor it is subjected to a barrage of neutrons. When the fuel ^{235}U absorbs a neutron it becomes a new isotope called ^{236}U, which is not stable. It will break up and decay. In fact there are many ways for it to decay, but a common one, illustrated in Figure 7.4, is;

$$^{235}U + n \longrightarrow\ ^{236}U \longrightarrow\ ^{92}Kr +\ ^{141}Ba + 3n + Energy$$

These reason energy is released is that the daughter particles, ^{92}Kr and ^{141}Ba, require fewer nuclear bonds to hold them together than ^{235}U. Again, a fuel is characterized by the fact that is starts out stable, needs a "spark" to ignite it, but then is reconfigured into something with weaker or fewer bonds, releasing the excess energy. This excess must be greater than the "spark," or in this case the number of neutrons, to sustain the burning.

What we are seeing is that potential energy that is stored in a material, a fuel, is closely related to the forces that bind that material (see Table 7.1). The amount of energy in nuclear fuel is related to the binding of neutrons and protons by the nuclear force. The amount of energy in other fuels, such as oil or food, is related to the binding of atoms by

Table 7.1 Comparison of nuclear and chemical forces and bonds.

Force/bond	Relative strength	Range	
Chemical	1	10^{-10} m	1 Å
Nuclear	10^6	10^{-15} m	1 fm

chemical forces or bonds. Now try to imagine an analogous system for gravity. Maybe planets with satellites as our molecule. If two planet-plus-satellite systems were to collide could they rearrange themselves and give off energy? Yes they could, but the amount of energy would depend on the particular planetary system and the "burning" of this type of gravity-fuel would be uneven and unsatisfactory even to a cosmic titan. There really is not a good analogous gravitational fuel.

The fact that nuclear fuels have about a million times the energy density of other fuels is because the forces are about a million times stronger. But this still is not a comparison at the most fundamental level. Chemical and nuclear forces are not basic and fundamental forces in nature. They arise as a local expression of deeper forces.

***

The basic four forces of nature are gravity, and the weak, electromagnetic and strong forces (see Table 7.2). In this list the nuclear force and chemical force do not even appear. That is because nuclear and chemical forces are only residuals. For example, a neutral atom is made up of electrons and a nucleus. These are held together by the electromagnetic force and the atom is neutral. So one would not expect an atom to be electrically attracted to another neutral atom, but they are. Atoms are attracted to each other because the charges inside of them are not

Table 7.2 Comparison of the four fundamental forces in nature.

Force	Relative strength	Range
Gravity	1	∞
Weak	10^{29}–10^{35}	$\sim 10^{-15}$ m
Electromagnetic	10^{36}	∞
Strong	10^{38}	$\sim 10^{-15}$ m

evenly distributed. So one atom's positive region may be attracted to a second atom's negative region and that attraction gives rise to chemical bonding. In a similar way the nuclear force between protons and neutrons is the leakage of the more basic strong force between quarks. We will describe this leakage in more detail in Chapter 10 for atoms and chemistry, and in Chapter 11 for the nucleus.

Right now, however, we will try to compare these fundamental forces to see if we can make some sense of the scales of energy. Is the strong/nuclear force intrinsically stronger than everything else? Is it really a million times stronger than the electromagnetic force? How weak is gravity?

The most common way of comparing forces is to find a place in nature where the two forces are at work and then compare their magnitudes in that situation. So to compare the strong and electromagnetic forces we can look at two quarks inside of a proton or neutron. If we do that we find that the strong force is about 50–100 times stronger than the electromagnetic force. There are two reasons that I can only give a range for comparing their strengths. Firstly, the ratio depends upon which quarks I pick; different quarks have different electric charges but the same strong binding. Secondly it depends upon the distance separating the quarks, and these forces act differently as distances grow. The electromagnetic force tails off over long distances, whereas the strong force grows as the quarks are pulled farther apart. In that respect the strong force is like a spring, harder to stretch the farther you pull it apart.

If the strong force is only a hundred times stronger than the electromagnetic force, why are the energies a million times different? This is because the constituents of an atom are almost 100,000 times farther apart than the constituents of a nucleus. The electromagnetic force drops off dramatically with this extra distance.

We can also use this technique to compare the strength of the electromagnetic force to gravity. This time we will use the simplest and most common atom to compare: hydrogen. The hydrogen atom is made of an electron and a proton. Electrons and protons are bound together primarily because they have a charge and therefore an electromagnetic attraction. They also have mass, and therefore exert a gravitational attraction. The electromagnetic attraction is 10^{36} times stronger than the gravitational attraction. In fact if you were to consult a number of references that compare forces, this is the ratio which is most often cited. This ratio is independent of separation distance, unlike our previous comparison, since the electromagnetic force and gravity tail off in exactly the same way at large distances. But the ratio is not without caveats.

The electromagnetic repulsion between two protons is only about 10^{33} times greater than the gravitational attraction. However, in the exotic and unstable positronium atom—made up of an electron and a positron (anti-electron)—the ratio is about 10^{39}. What is happening here is that in all three cases the electromagnetic attraction or repulsion is exactly the same, but gravity depends upon the mass of the particle involved. The reason why I have offered the positronium and repelling proton case is that it reminds us that the commonly quoted ratio of 10^{36} is not an absolute of nature. It is a function of the system we picked to use when comparing the forces. The comparison of forces may be instructive, but it is not absolute.

So here we are at the end of our discussion of energy. We were trying to find some large guiding principle such that we could order small energy objects at one end of our scale and large energy objects at the other end. We then saw a separation between chemical fuels and nuclear fuels, but energy related to gravity did not fit. We also saw that chemical and nuclear forces are not fundamental, but rather forces derived from the residues of the electromagnetic and strong forces. There are trends here, but still no deep enlightenment.

There is one force I have not discussed much: the *weak force*. This is responsible for radioactive decays. Like the strong force it is short ranged. Also, as the name implies, it is very weak: about 1–7 orders of magnitude weaker than the electromagnetic force. The gravitational force may be trivial in strength, but it does have a long reach, all the way across the universe. The strong force might be short ranged, but it is powerful. The weak force is both short ranged and feeble. It also does not contribute to the shape of anything in the universe and so it is tempting to neglect it. But the weak force lies behind radioactive decay and if there was never a radioactive decay much of the energy tied up in the nuclear force would never be released. The weak force is what unbalances the systems of quarks, neutrons and protons and sets the pace for the release of nuclear energy. It is not only the reason that stars burn, but why they burn over eons and not instantly. It can be argued that the universe has a history of 10^{17} seconds and is still dynamic and still unfolding because of this force.

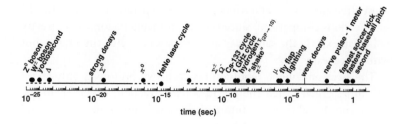

time (sec)

❧ 8 ❧

Fleeting Moments of Time

I am sitting on the grass of Salisbury Plain staring at Stonehenge and thinking about time. When you are sitting inside of earthworks that are 5000 years old, looking at stones that were dragged here as much as 4500 years ago, it is hard to not think about time. It is tempting to think about this stone ring as a hole in the fabric of the universe, a portal into another age, an era before the machines of today. Can I see a bit of the Neolithic world? Or perhaps it is the other way around and it is a bit of the Neolithic world protruding into our own age? It may be tantalizing to think about Stonehenge in those terms, even picturesque and poetic, but it really is not quite right.

Those blue stones were raised up on end about 4500 years ago and they have passed through all 4500 intervening years to get to the present. They are not in the Neolithic world any longer and they bear the scars of age to prove it. I am sitting on the plains of Salisbury in the twenty-first century on a beautiful sunny August day. The forests of the Mesolithic and Neolithic have vanished and instead there are sheep grazing on the downs. I am in the midst of a crowd of a hundred people speaking a dozen languages and coming from six continents. The stones themselves are also travelers, much older than a mere 4500 years. Geologists tell us that the blue stones of Stonehenge came from the Preseli Hills in Pembrokeshire in the southwest corner of Wales. They were formed in the Ordovician era, about 450–500 million years ago.

So how old is what I am looking at? These stones and earthworks were arranged here between four and five thousand years ago. But it was in a volcanic eruption 100,000 times older than Stonehenge that the blue stones were formed. And the atoms and quarks within the stones? They date back to a time shortly after the big bang, 13–14 billion years ago.

The formation of the stony Welsh mountains, the building of Stonehenge, this day when I sit in the sunshine on the grass of Salisbury Plain are all "events." It is time that separates events that take place in the same location. I may be at Stonehenge, but I am not witnessing John Aubrey performing his seventeenth century survey, or the Battle of the Beanfield of 1985. I am not bumping into the Neolithic engineer who paused at this spot to consider the stone's alignment. For we are separated by 4500 years.

***

Over two chapters I will describe the many scales of time: in this chapter short time and in the next chapter long time. We will look at the types of events that are measured in years or seconds, in eons, or in the "shakes of a lamb's tail." We will also look at how we measure time and how we would recognize a good clock. Also how we can date the universe, measure the lifetime of an exotic, fleeting particle, or measure the speed of a nerve pulse.

***

Time seems like such a simple thing as it flows by us, or as we pass through it, from the past into the future. We like to think of time as a line, and even chart history on a timeline, with the Romans a long way off to the left, the Middle Ages closer and the present right in front of us. And off to the right? That is where the question mark goes; that is the unknown future. So we can think of time as just a line, like latitude or longitude, that we travel along. In fact, in relativity, we are taught to calculate the separation of events by combining latitude, longitude and elevation (x, y, z) with time (t) in an equation that looks like the Pythagorean theorem for triangles. However, time is not just like distance in a different direction. We are always compelled to travel along it. We cannot sit temporally still and stay in the same moment, and we most certainly cannot move backwards. That makes time a unique type of dimension. We must follow *time's arrow* in one direction, towards the future.

For millennia, people have written, discussed, and argued about what time really is and the nature of this thing called time's arrow? Is it a necessary feature of the universe? Is it a consequence of entropy and the second law of thermodynamics? Or is it the other way around? The philosophy of what time is is beyond the scope of this book. Instead I will pose two simpler questions that we can answer. Why do we know time exists? And how do we measure it?

We know time exists because we see change. That may seem self-evident, but not to Parmenides of Elea. Parmenides (520–450 BC) could not understand how things could move, because in his view the universe was packed tight. He rejected the idea of empty space and is attributed with saying "Nature abhors a vacuum." In his view the world is tightly packed and so you have no space to move into. Therefore motion is an illusion and time is unreal. His most famous student, Zeno, left behind a number of motion paradoxes to puzzle us. I will, however, take the more pragmatic view that motion is not an illusion, for if it was, trying to do anything, such as write a book, would be pointless.

***

We know time exists because we see change, which also gives us a handle on how to approach the second question: how do we measure time? To measure time we look at how something is changing. For example, I could mark time by the growth of a tree or the march of a glacier. Since the last time I visited these mountains it has been 25 tree-millimeters or 5 glacial-meters. But these are not the best candidates for a clock because on my next return I may find only 22 tree-millimeters but 8 glacial-meters of time, meaning these two timepieces are out of synch. What I want for a good clock is something with consistent motion or constant change.

Humans have used the Sun, the Moon and the stars to mark time since before our collective memory. The day, as marked by the passing of the Sun, is ingrained in our daily routine. We rise and set with the Sun. But we needed to subdivide the day into smaller chunks of time—into hours—and the most natural tool for doing this is the sundial. The shadow cast by a stick, or *gnomon*, sweeps across the ground as the day progresses. But the rate of that sweep is not uniform. It creeps at mid-day and rushes when the Sun is on the horizon. The difficulty arises from the fact that the Sun follows a circular path—an arc—across the sky whereas the shadow is confined to a flat plane. Not only is there this arc-versus-plane mismatch, but the arc is tilted at an angle that depends upon your latitude on the globe as well as the season.

Through good design you can solve part of this problem. The sundial face need not be a flat plane. If it too is an arc then the shadow will march an equal angle and distance for every hour. Also you can tip the dial to compensate for your longitude and in the process create a beautiful, useful and accurate time piece . . . almost. The most obvious problem is what do you do the other half of the time: at night?

Determining the hours through the night can be done with something called an *astrolabe*. This essentially measures the positions of stars and converts that information to hours, much like a sundial measures the position of the Sun. An astrolabe needs to be a bit more sophisticated than a sundial because the night sky changes throughout the year. For example, Orion is a winter constellation whereas Scorpio is most visible in the summer sky. However, since the march of the constellations across the sky is very regular an astrolabe has a disk that you turn to the date, and then you sight one of many stars that are visible and read off the hour.

Cloudy days call for a different solution, but humans have been resourceful and have solved this problem with clocks based on dripping water, burning candles and flowing sand. In all these cases we knew that we had a good and regular clock if it exhibited a constant motion or rate of burn, especially when compared to a sundial or astrolabe. A good clock is one that agrees with the Sun.

As our pendulum clocks became more and more accurate a problem in our very concept of a day became more and more apparent. With a pendulum clock we could compare the hours of a sundial and astrolabe and find that they did not agree; that is, the Sun and the stars disagreed on the length of a day. To understand this problem we need to step back and think about what we mean by a day. A day is the amount of time it takes for the Sun to pass due south of you (if you are in the northern hemisphere) until the next time it passes due south of you. Now let me build my best pendulum clock and adjust the length of the pendulum such that it ticks once a second. A day contains 24 hours, each hour has 60 minutes and each minute has 60 seconds, so there are 86,400 seconds in a day. Now in the middle of July I meticulously adjust my pendulum such that from noon to noon my pendulum swings pack and forth 86,400 times. I can even cross check that with the stars. I can count the ticks from the time when Altair, a star in the constellation Aquila, crosses my north–south meridian one night until it does so again the next night. I count 86,164 seconds. This is okay because stars mark out *sidereal time* and the Sun *solar time* and they shift by about four minutes a day because we are going around the Sun. Four minutes is 24 hours divided by 365 days. This is the same reason that different constellations show up in the summer and winter sky.

To prove that our clock is a good one we will repeat our measurement at the beginning of the New Year. Altair is not visible at this time of year, but we can use Betelgeuse in the constellation of Orion. The time for Betelgeuse to transit the meridian from one night to the next is 86,164 seconds. The sidereal day is the same as when I measured it in July. But my measurement of the Sun from noon to noon is about 86,370 seconds! The day is half a minute shorter than we expected.

Sundials gain time from February to May and again from August to November. They lose a little bit of time from May to August and a lot of time from November to February; as much as half a minute a day. The reason for this is that the axis of the Earth is tilted compared to the plane of our orbit and the fact that our orbit is not circular; it is an ellipse with our motion speeding up as we approach perihelion (January 3), the point where we are closest to the Sun. That means that this time creep is very regular and we can adjust for it. The adjustment is called *the equation of time*. Sometimes it is combined with the north–south seasonal changes to give a figure-of-eight figure called the *analemma* (see Figure 8.1).

These adjustments are calculable, but they leave one with a dissatisfied feeling. Whenever we have a new technique for measuring time we find that it is better than previous techniques, but it never seems to be perfect. Further corrections and adjustments always seems to linger. We are left wondering if we always will need to tweak and fine-tune our clocks.

The stars really are a good way of marking time. They travel across the sky at a very regular pace because the Earth is rotating at a very regular rate. Our home planet is a ball spinning in frictionless space and, by Newton's laws and the conservation of angular momentum, we will keep spinning at the same rate far into the future. The same stars will cross the meridian every 23 hours, 56 minutes and 4.0916 seconds . . . almost . . . if it were not for the tides.

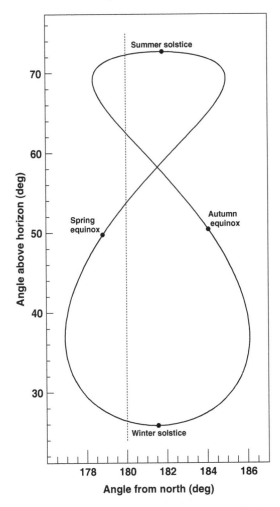

Figure 8.1 The *analemma*, as viewed from New York City. The location of the Sun at 12:00 noon changes through the year: north–south due to the season, east–west due to the shape of our orbit. The dotted vertical line is due south. It is not centered since New York City is not in the middle of its time zone.

The Earth's rotation rate is actually slowing down and the speed at which the Moon orbits is speeding up. It is not changing much. Our days will be a second longer 40,000 years from now and if our Sun were to burn long enough the Moon's orbit and our rotation would eventually synchronize. A lunar month would equal a solar day. This is because of tides. We think of tides as a seaside phenomena, but the Moon pulls on the waters mid-ocean as well. In fact the Moon also pulls on the rocks in our planet, flexing and distorting our crust twice a day, each and every day. This flexing of rocks from the tidal effect works like friction, slowing down our rotation and every century adding 2.3 ms to each day. But tides are not the only thing affecting the Earth's rotation rate.

The Earth is changing its shape and has been since the end of the ice age. With glaciers retreating, continents, especially near the poles, have undergone *rebound* and are rising up. This changes the rate of our rotation in the same way a spinning ice skater can change their spin rate by pulling in their arms. This speeds our day up by −0.6 ms per day per century. Tidal and post-glacial rebound combine to give a slowing of 1.7 ms per day per century. That does not sound like much, but it adds up. The year 1900, which was taken as the standard year, is now a century ago, which means our days are now 1.9 ms longer than in 1900, and by 2100 they will be 3.4 ms longer. Presently that 1.9 ms adds up to a whole second after about 526 days at which point we need to add a *leap second* to our clocks. The rate of correction will only increase.

In addition there are other things going on that have less predictable effects. For example the Indonesian earthquake in 2004, which led to a tsunami, shifted enough mass under the Indian Ocean to effect the Earth's rotation rate by 2.68 μs. When the winds pick up for El Niño, the Earth slows down. When ocean levels change and their mass moves, these changes are measurable, but have somewhat erratic effects on the length of day. This is why the adding of leap seconds to our clocks is sporadic. It starts to make one think that the rotation rate of the Earth might not be the best way, the ultimate way, of measuring time.

We need to go back to the question of what does it mean to say that we have a good clock? It means that when this clock marks out a day or a second it is the same amount of time as the next day or second that it marks. But that does not tell us how we know. The way we know that a clock design is good is to build a bunch of independent clocks, start them together and see how they vary after a while. If we have ten clocks of one design and after a day they vary by an hour, that is not a good design. However if after a year they very by a tenth of second, that is a much better timepiece.

Through the centuries, horology, the study of the measurement of time, has devised a number of techniques for building clocks. These include the chronometers used to solve the longitude problem of navigation. Every generation created more and more accurate clocks. Today the gold standard of horology is a type of atomic clock called the *cesium-fountain clock*. The title of the best clock is a coveted prize and is constantly shifting between various national standards institutions and observatories throughout the world. In fact, we can only determine how good a clock is by comparing it to other accurate clocks. In the US, there are cesium-fountain clocks at the Naval Observatory and the National Institute of Standards and Technology. In Germany there is one at the *Physikalisch-Technische Bundesanstalt*. In the UK the National Physical Laboratory hosts NPL-CsF2, which at this time is the best clock on Earth, accurate to 2.3×10^{-16} or 10^{-11} seconds a day. Another way of think about that is if you had two clocks like NPL-CsF2 and set them going, after about 100 million years they would differ by no more than about a second. If you had started these clocks at the time of the Big Bang, they would now differ by no more than two minutes.

Time is something that we can measure very well because we can isolate atoms and get them to do the same atomic transition again and again in a very repeatable fashion. The heart of a cesium clock is the transition between two atomic states in cesium-133 atoms. Light from this transition has a unique frequency and when that light oscillates 9,192,631,770 times, that is our modern definition of one second. The challenge in accurate clocks has been to separate those cesium atoms from external effects. So the "fountain" in the clock's name is a way of letting the atoms free-fall, to remove gravitational effects. They are also cooled with lasers to remove thermal jitter.

With an accuracy of 10^{-16} these clocks represent our best measurement of any type, better than length or mass or energy. The second is exactly 9,192,631,770 oscillations and a day is defined at 86,400 of these cesium-based seconds. It is close to the time for the Earth to rotate, but no longer does the rotation rate define time.

With time measured so much better than anything else it also gives us a new way of setting distance standards. In fact we now define the speed of light by saying that in one second light travels 299,792,458 meters *exactly*. We will never add more digits to that number. It is 299,792,458.000,000. That means that the meter is the distance light can travel in 1/299,792,458 s. This also matches that ideal that the original designers of the metric system had: anyone anywhere with a good clock can create a standard meter.

***

Now let us take our clocks and start to measure our world. We will start by looking at things that take about a second and then look at the types of events that take even less time. In the next chapter we will look at timescales greater than a second.

As I said at the beginning of this book we can think of a second as being about the time between heartbeats. It is also about the time it takes for an old-fashioned clock to "tick-tock." In playground games it is the time it takes to say "one thousand" or "Mississippi."

A bit faster then a second is the timescale of sports. One of the fastest baseball pitches ever recorded was thrown by Aroldis Chapman of the Cincinnati Reds, clocked at 105 mph (46.9 m/s). Since it is 60 feet 6 inches (18.44 m) from the pitcher's mound to home plate that means the ball took 0.39 s to get there.

Soccer goalkeepers have even a greater challenge. In a penalty kick, the ball is typically moving at about 30 m/s over a distance of 11 m. So the goalkeeper has about 0.36 s to respond. One of the fastest soccer balls recorded was nailed by David Hirst, clocked at 114 mph (51 m/s) in 1996. He was playing for Sheffield Wednesday against Arsenal. The shot was taken from about 15 m out, which means the goalkeeper had 0.22 s to respond. Unfortunately for Sheffield Wednesday the ball hit the cross bar and bounced out.

What makes most ball-based sports exciting is this matching of human response time to the motion of the ball. Human response time for young adults is about 0.2 s, which is well matched to sports. If all shots on goal were from 30 m out, or pitchers threw from second base, goalkeepers and batters would almost always intercept the ball. If much closer, they could rarely stop it.

***

A bit faster than sports times are the frame rates of film and video. The brain can process about a dozen images a second and so if it is presented with more than that an illusion of continuous motion is created. Some earlier movies were filmed at 14 frames per second (FPS) but, over time, the frame rate has generally increased. Presently most films and videos are 24, 25, or 30 FPS, but there is a push to increase this. A lot of sports are recorded at much higher rates, so that they can be played back in slow motion. Also, modern screens and projectors are more versatile than projectors of a century ago, so these numbers can easily change. Still, 25 FPS is a time of 0.04 s per image.

Nerve pulses in mammals move at a speed of about 100 m/s. The longest human nerve (the foot to the center of the back) is about 1 m long, which means a signal will travel the length of a nerve in about 0.01 s. By modern standards that may not seem like such a short period of time, but before Hermann von Helmholtz (1821–1894) made his measurement in 1850 it was thought to be so fast that it was not even worth trying to measure. What Helmholtz did was to take frog's legs with the nerves attached and exposed and mounted them next to a swiftly rotating drum. The legs were attached to a stylus such that when they twitched they left a mark on the drum. Helmholtz then stimulated the nerves. By looking at the marks from the stylus on the drum, and how far the drum had rotated, he determined the muscle's response time. He then changed the point of stimulation on the nerves and was able to measure a nerve pulse propagation speed of about 30 m/s. In mammals, nerves are a bit different than in frogs and the speed can be up to 120 m/s.

Track events are now recorded to the hundredth of a second. Actually they are timed at an even faster level, but then rounded off and reported to the nearest hundredth. To get the times that accurate requires electronic starts and stops, but the human eye can actually see the difference. For the one hundred meter dash at Olympic speeds, one hundredth of a second is the same as 10 cm, a distinction a sharped-eyed finish-line judge can see.

A thousandth of a second is called a millisecond (ms). It is about the time is takes a nerve pulse to jump across the *synaptic gap* between nerves. A fly can flap its wing in about 3 ms.

***

Lightning is a spark of electricity that passes between the Earth and a thundercloud in about 10^{-5} s or 10 μs. Actually that is just the central step of lightning. Before a strike a *leader* will form, a region of ionized air that is more conducive to lightning. That formation can take tenths of a second. The discharge itself is really too fast for us to see, but in the process the electric discharge superheats the air. What we see is hot glowing air, which lingers after the discharge itself.

The muon, a subatomic particle that acts like a heavy electron lives for about 2 μs (2×10^{-6} s) if it is standing still. In fact the lifetime of the muon was one of the first experiments to test special relativity. If a large number of muons were traveling at nearly the speed of light, and there were no relativistic effects, half of them would decay in about 600 m. So experimenters counted the number of muons streaming down from the sky, first on a mountain top. Then they repeated their measurement at a lower elevation and found that far less than half of them had decayed after 600 meters. This was exactly what special relativity had predicted. The lifetime of a particle is according to its own measurement of time, and when something is in motion, especially as it approaches the speed of light, its internal clock slows down. According to the muon, it had traveled that 600 m in less than 2 μs (time dilation) or, equivalently, the distance it traveled had shrunk (length contraction).

Relativity also affects very accurate clocks that travel at much slower rates. A GPS satellite travels at about 14,000 km/h so its clocks will be slowed by about 7 μs a day by special relativity. Actually there is a larger effect due to the way satellites travel through the Earth's gravitational field. General relativity speeds up the satellites' clocks by about 45 μs a day. If these were not corrected, the position of the satellites would be off by a few kilometers a day, getting worse as time went on.

Quicker than a microsecond is a nanosecond (ns), a billionth of a second. *Gigahertz* means a billion cycles per second, so a 2-GHz computer has a clock generating two ticks per nanosecond. Light can travel 30 cm, or about a foot, in a nanosecond. Electrical pulses in a wire are slower than light, which means that if I had a computer with a 10-GHz clock, two pulses would be separated by about the width of a chip, which must make a very interesting design challenge.

During the building of the first atomic bomb the term *shake* was coined to refer to 10 ns. This term was inspired by the phrase "two shakes of a lamb's tail," but actually referred to the amount of time between steps of a nuclear chain reaction. In the previous chapter we briefly described the fission of a uranium nucleus into krypton, barium and three neutrons. Those neutrons may travel a few centimeters before they are absorbed into another nucleus to start the next step of the chain reaction. It takes about 10 ns, or one shake for that to happen. Depending upon the particular fuel and the other materials in the reactor core or bomb, the designer can vary what this time is a little bit.

As we look at faster and faster events we are approaching the intrinsic speed of our clocks. The standard transition in cesium-133 produces photons with a frequency of 9,192,631,770 Hz, which means a wave passes by in 10^{-10} s. At about 3 cm this is actually a pretty long wave. We could have used a different transition. For example, the red light from a helium–neon (HeNe) laser has a wavelength of 633 nm and a frequency of 4.7×10^{14} Hz, so a wave goes by every 2×10^{-15} s. The reason we do not use these for clocks is because they are not as stable or as reproducible as the cesium-133 transition. However, this uncertainty in faster atomic transitions is something we can use to push our measurements to even shorter times.

***

One of the most discussed, most mysterious and most misunderstood concepts in quantum mechanics is the *Heisenberg uncertainty principle*. It is a relationship between how well you can know two related quantities such as the position and momentum of a particle. Its most infamous misuse is to declare the unknowability of something: "according to Heisenberg, I can't know anything." What Heisenberg really tells us is that if we measure one quantity of a particle with great accuracy and rigor we will have disturbed the system so much that the complementary quantity can only be measured with limited accuracy. So if you measure position well it affects momentum; if you measure energy well it affects the time of the event.

For example, imagine you want to measure the energy of a ball that has been kicked or thrown. You could hang a block of foam from a rope in front of the ball. When the ball hits the block, the block is swung back. The swing is the measurement; the more it swings, the more energy the ball had. But in the process you most certainly have affected the ball. It would now be hard to answer the question of how far the ball would have traveled if the block was not there. You can try to come up with better methods of measuring energy; a smaller block will effect the ball less, or perhaps you could just scatter light off the ball, which is essentially what a radar gun does. But even these more benign techniques will still affect the trajectory of the ball.

This is where Heisenberg started, and his principle transcends any particular apparatus or technique and talks about interactions at the quantum-particle and wave levels. The most familiar form of the uncertainty principle is a relationship between position and momentum.

But the princple also says that there is a relationship between energy and time. The uncertainty in energy multiplied by the uncertainty in time is greater than or equal to Planck's constant.

$$\Delta E \; \Delta t \geq \hbar$$

Planck's constant is a very small number so this relationship rarely impedes laboratory measurements. But it is not just a limit of experiments designed by humans. It is also a limit on measurements nature can make upon itself. This may seem odd, but it is true. "Measurement" sounds like such a deliberate act, but in fact measurements are happening all of the time. If a cue ball hits another ball on a pool table, the second ball goes off with a well-determined speed and direction. The second ball's momentum is set because it "measured" the momentum lost by the cue ball. An interaction or a collision is a measurement, and therefore Heisenberg's principle is relevant.

An application of the uncertainty principle is to look at the width of a spectrum line. An atomic spectrum usually looks like a black band with some thin strips of different colors on it. The way it was created was that a gas was heated or sparked until it glowed. The light from the glow was then passed through a prism or grating that separated the colors into a rainbow with red at one end, green in the middle and violet at the other end. If we have a simple and pure gas and a good set of optics we will see distinct bright lines.

For example, if we use hydrogen we will see four lines: one red, two blue and one purple. There is also an ultraviolet line that we do not see, as well as lines in the infrared. Photons that contribute to that red line were created when the electron in a hydrogen atom was excited into the third orbit and then fell back into the second orbit. The energy lost in that fall from the third to the second orbit was converted to a photon and that energy determined its wavelength and hence the fact that it is red. But if we look closely at that spectrum line we see that it is not a perfect mathematical line; rather it is a band with a width. If we look across that band the hue of the red slightly changes. That is because the photons that make this band have slightly different energies: there is a spread or uncertainty in their energies. This spread is exactly related to the amount of time the electrons spent in that excited state, that third orbit. The longer an electron dwells in an excited orbit, the more time it has had to settle, the more precise its energy. So we can use the spread in energy or color to calculate the time spent in the excited orbit. By looking at line width, we measure time.

A well-studied spectral line is the one related to the jump between the first and second orbit of the hydrogen atom. We do not see this one because it is in the ultraviolet. The jump itself has an energy of 10 eV. The line itself is exceedingly narrow at only 4×10^{-7} eV. The time related to this line is a bit more than a nanosecond:

$$\tau_{2P \to 1S} = 1.6 \times 10^{-9} \text{ s}$$

In fact most atomic transitions are measured in nanoseconds or slower.

★★★

If we want to find events in nature that happen faster than atomic transitions we need to look at things that are smaller than atoms. So we will look at the decay of exotic particles. In Table 8.1 I include a very short list of particles and their lifetimes. It is by no means exhaustive. I have selected these because they show trends.

The pions (π^+, π^-, π^0) are related to the nuclear force. The sigma particles ($\Sigma^+, \Sigma^-, \Sigma^0$) are much like protons or neutrons except they have a *strange-quark* in them. The omega particle (Ω^-) is made of just strange-quarks. The delta particles ($\Delta^{--} \Delta^- \Delta^0 \Delta^+$) are like neutrons or protons with complex spins. We will talk more about these later.

All three pions are nearly the same in mass and structure, but π^0 decays a billion times faster than the other two. All three sigma particles are nearly the same in mass and structure, but Σ^0 decays a billion times faster. Clearly there is something special going on. A billion is a huge

Table 8.1 Exotic subnuclear particles; lifetimes, forces and decay.

Particle	Lifetime (sec)	Decay force
$\pi^+ \pi^-$	2.6×10^{-8}	weak
π^0	8.5×10^{-17}	strong
$\Sigma^+ \Sigma^-$	8.0×10^{-11}	weak
Σ^0	7.4×10^{-20}	strong
Ω^-	8.2×10^{-11}	weak
$\Delta^{--} \Delta^- \Delta^0 \Delta^+$	5.0×10^{-24}	strong

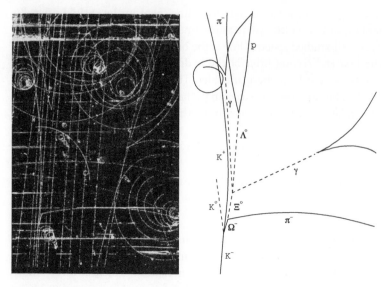

Figure 8.2 The discovery of the omega particle. Left: a photograph of a bubble chamber and particle tracks. An accelerator creates a spray of various subnuclear particles. When the particles pass through the chamber they leave a trail of bubbles that we can see. Right: Each track is identified as a particular type of particle. Courtesy of Brookhaven National Laboratory, 1964.

factor when comparing the lifetimes of such similar particles. It would be as if a germ that lived an hour had a cousin that lived 100,000 years. That is a factor of a billion.

A picture is said to be worth a thousand words, so I am including a photograph of the discovery of the omega particle (see Figure 8.2). On the left is a photograph taken at Brookhaven National Laboratory. The white lines are really a line of bubbles left by particles passing through liquid hydrogen. On the right are the interesting tracks with the particle names. The whole image is about half a meter wide. Because the energy and speed of each particle was known, its lifetime could be determined. That omega particle traveled only a few centimeters, which translates to a lifetime of about 10^{-10} seconds. In this way we can measure the lifetime of fast-moving, but short-lived particles. Also relativity helps. Much like when we talked about muons a few pages ago, fast-moving particles live longer and travel farther.

But why do π^0 and Σ^0 decay quickly when compared to their siblings? The reason is that these decays are due to the strong force, whereas the $\pi^\pm$ and $\Sigma^\pm$ decay due to the weak force. We touched upon these forces briefly in the last chapter on energy. Weak decays are the transition that set the burn rate for hydrogen in the Sun, as well as the decay of neutrons. They set the reaction times for both nuclear fission and fusion. And as we can see in the table above, weak decay is a billion times slower than strong decay. In fact it was this factor of a billion that helped people identify that the weak and strong forces are unique and distinct types of interaction.

Once we identify that the $\Sigma^\pm$ is undergoing a weak decay in which a strange quark is decaying, it is then easy to recognize that this is also what is happening inside of the omega particle. However, the delta particle is something else.

***

We often think of a proton or a neutron as a solid, hard sphere in the center of an atom. We are use to the idea that an atom can be excited and we know that, related to that excitation, atoms can radiate light. But the idea that a proton or neutron might also be excited seems foreign. But that is what a delta particle is.

Inside of a proton or neutron there are quarks that travel around each other, much as electrons swirl about the nucleus of the atom. The quarks are confined to an area about 1×10^{-15} m across. That is an region 100,000 times smaller than an atom, and the energies are about a billion times greater. Each quark has a number of properties that affect how it interacts with the other quarks. For the moment I want to focus only on one property called *spin*. Spin interactions are like the interactions between bar magnets. If I have two magnets next to each other they would like to align themselves with opposite poles together: opposites attract. In a neutron or a proton one quark will have its spin opposite the other two, to minimize this magnetic tension. If you twist around the spin direction of that quark until they are all aligned you need to add a lot of energy. You have effectively "excited" that neutron or proton, creating a new particle called the delta.

You can also twist the spins of electrons in an atom and in the process add a tiny amount of energy, but when you do this with quarks the effects are huge. You need to add a few hundred megaelectronvolts of energy. For comparison, the mass of a nucleon is 940 MeV. This means you need a fairly large accelerator to do this. The delta particle is not very stable. It exists for a few short moments and then decays back into a proton or neutron, kicking out the added energy in the form of a new particle. If we look at the energy of that ejected particle we see that it is on average 300 MeV, which means the delta is 30% more massive than the original proton or neutron. But not all of the ejected particles have energy of exactly 300 MeV. In fact they have a distribution, centered at 300 MeV and with a width of 120 MeV. We can use Heisenberg's uncertainty principle and convert this into time and find that the lifetime of the delta particle is 5.6×10^{-24} s.

A time of 10^{-24} s is also called a *yoctosecond*. It takes light about half a yoctosecond (ys) to cross the diameter of a proton and so it is not surprising that this is the timescale for events inside of protons and neutrons. The W boson has a lifetime of 0.3 ys and the Z boson has a lifetime of about 0.25 ys. We have not seen anything significantly faster than this. Will we some day? I do not know. But as I said before, a yoctosecond is about the time it takes for light to cross a proton. That is not a coincidence. For a proton to be excited the whole structure of the proton is involved and so the time is related to how long it takes light to cross that structure. If someday we find smaller structures, perhaps something inside of quarks, then I would not be surprised to find events that are even faster.

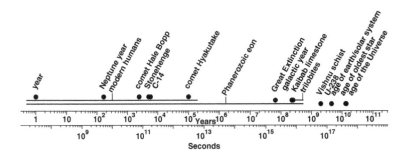

Deep and Epic Time

In the previous chapter we talked about the second, the time of a heart-beat. We said it was a chunk of time in which a cesium-133 atom could undergo a transition with a resonance of 9,192,631,770 Hz. We also talked about events that take less than a second: lightning strikes, nerve pulses, atomic transitions and the lifetime of subatomic particles. But we do not measure our lifetime in seconds. We measure events of our lives in terms of days, seasons, years, decades and even centuries. The basic SI (Système International, i.e. the metric system) unit of time is the second, but these other units have real value. The year, the week and the hour are directly tied to our experiences and it is hard to imagine functioning in a world without them. But they are a human construct, and as such we could have picked some other division of time, and in fact in the past we have.

The Chinese historically divided the day in various ways. Sometimes they divided the day into twelve *shichen*, a double long hour. Alternatively they divided the day into one hundred *ke*. In this sense, the *ke* is a type of decimal time. But the *ke* and the *chichen* did not compliment each other well. So at various times in Chinese history, the *ke* was re-defined as 96, 120 and 108 per day. There were also a number of shorter time units such as the *fen* and *miao*, which are similar to the minute and second.

Hindu time is very complex, and spans from epic times, measured in trillions of years, to parts of seconds.

In the Middle Aages the church developed eight *canonical hours*: matins, lauds, prime, terce, sext, none, vespers and compline. These were times in the day to stop and pray, particularly important in monastic life. They were based on a Judaic tradition and are much like the five daily prayers in the Islamic world today. But they are not really "hours" in the sense we use that word today. Rather, they are events within the day.

In Europe, the day has been divided into twenty-four hours for centuries with a few interruptions. The most notable exception was during the French Revolution. I am going to describe the clock and calendar of the early French Revolution so we can see how time may have been measured, as well as understand why it is not measured that way today.

In 1793 *decimal time* was introduced into France. In this system the day was divided into ten hours, and those hours were subdivided into minutes and seconds. Since this system re-used the words, *hour, minute* and *second* it would get confusing when I compare them to our standard units, so I will refer to them as *decimal hours, decimal minutes* and *decimal seconds*. One day had ten decimal hours, one decimal hour had one hundred decimal minutes and each decimal minute had one hundred decimal seconds. Since the decimal and the standard time units are based on the common day we can compare them. What jumps out at me in Table 9.1 is that these times are not so radically different from what we are use to, especially the second. If your heart beats 70 beats in a standard minute, a fairly normal rate, then it beats once every decimal second.

Decimal time was the official way of recording time in the young French Republic and it was closely related to the French Republican Calendar or *revolutionary calendar*. In this calendar every week, or *décade*,

Table 9.1 Comparison of decimal and standard time.

Decimal time	Standard time
1 decimal hour	2.4 standard hours
1 decimal minutes	1.44 standard minutes
1 decimal second	0.864 standard seconds

had ten days and every month had three weeks. Under this system a year—one trip of the Earth around the Sun—would take 12 months, five and one quarter days (or 2.5 decimal hours). The year is still an important unit of time and even the Enlightenment could not change when we should put seeds in the ground to grow things.

The calendar was not an easy sell. In it there are only 32 *décades* (weeks) each year and each *décade* only had one day of rest. To make it more palatable a mid-*décade* half-day holiday was added, and those five or six days that did not fit into the twelve months were declared "complementary days," or national holidays, to be taken in the middle of our September.

The reason the French revolutionary leaders went to such lengths was to try and break the power of the church. The revolution was a secular movement; it was a rising up against not only the monarchy, but also against the Church, which authorized and endorsed the king and court. In the eyes of the revolution, the Gregorian calendar—the calendar filled with the days of the saints—as well as Sunday—the day of worship—together represented a stranglehold that the Church held on society. By wiping the calendar clean, they tried to weaken the hold of the Church.

The revolutionary calendar was colorful. Since days were no longer named after saints, they needed new names and the gardens of France provided them, for example *Hyacinthe* (hyacinth) on April 28 and *Fraise* (strawberry) on May 30. The complementary days and holidays were even more colorful: there were fetes *de la Vertu, du Génie, du Travail, de l'Opinion* and *La Fête des Récompenses*. Finally leap day became *La Fête de la Révolution*.

The decimal calendar and clock were adopted in 1793 but did not survive. The clock was discarded in 1795 and the calendar in 1806. On the same day that the decimal clock was discontinued, the metric system, with its meter and kilogram, was adopted as the standard of the new republic. This leads us to the question, why has the meter thrived whereas decimal time has vanished? The main reason is that the meter unified commerce and decimal time did not. Prior to the introduction of the meter there were dozens of length standards, most of them unique to a local market. It was frustrating to a merchant who would buy, for example, a length of rope in one market and sell it as a shorter rope in the next city. It was also paralyzing to a tax collector. How could you enforce a uniform law or regulation in which lengths and weights had no uniformity? So the meter and the kilogram were welcomed with open arms, especially by people who traveled between markets. Decimal time, however, did not offer this advantage. Time was already uniform across France and most of Europe. Ironically the watch and clockmakers of France lost business. Before the revolution watch- and clockmakers such as Abraham-Louis Breguet were setting the standards for the rest of the world, creating state-of-the-art timepieces. However the ten-hour watch was provincial and barely sold even within France. Decimal time was a failure because it did not unify.

***

No matter what system is used, a clock and calendar still need to be synchronized with the rotation rate of the Earth and the time to orbit the Sun. Days and years are very real. So, we have ended up measuring time in the most curious set of units. Sixty seconds is a minute and sixty minutes make an hour. Then twenty-four hours make a day. But why 60 and 24? We inherited these numbers from the Babylonians and Sumerians who were using base sixty more than 4000 years ago. But why has its use persisted? In part it is because there seems to not be any overwhelming reason to change a system that gets the job done and is universally understood. But also it is very easy to divide these two numbers into a lot of factors. Sixty can be divided into 2, 3, 4, 5, 6, 10, 12, 15, 20 and 30, which is a pretty impressive list. It is the smallest *unitary perfect number*, which may not be the reason it is chosen, but instead is related to the fact that you can spend a quarter or a third of an hour on a task, which is harder to do in decimal time. Twenty-four is also no slouch, being divisible by 2, 3, 4, 6 and 12, all of which are handy numbers when trying to divide up the day into shifts or watches.

When you look at the success of the 24-hour clock and the metric system, it makes one realize that the meter has become the world standard not just because the conversion between centimeter and kilometer is easier than inches to miles, but because the meter replaced confusion, whereas decimal time did not.

***

Now we will look at events that happen on ever-increasing timescales, starting with events that take a few minutes and working our way up to the age of the universe. To be consistent with the previous chapter I should label all times in seconds, but most of us do not have a real feel for how much time a million seconds represents. So I will end up using a mixture of common units.

What are events in nature that take more than a second? The average lifetime of a neutron is 880.1 s or 14 min 40 s. This number comes with a lot of caveats. A neutron bound into the nucleus of hydrogen or carbon-12, or in fact any stable isotope can sit quietly through the ages, stable and not decaying. But pluck that neutron out of that nucleus, remove it from its safe haven, its nest, and in less than a quarter of an hour it will decay. For the moment I leave this observation as just a curiosity.

Under ideal conditions, some bacteria can divide in 20 min, a potentially frightful prospect if it is a pathogen and when you think about exponential growth. The eruption of Stromboli happens a few times an hour. Old Faithful, the geyser in Yellowstone, erupts every 45–125 min. All these events happen frequently, but not with perfect regularity. As we look for events with longer and longer time periods we do not find real uniformity until the 24-hour day.

One rotation of the Earth about its axis: a day, 24 hours or 86,400 seconds. It is one of the most significant chunks of time in human existence. I can imagine a year of 500 days, or even living in a place without seasons and so effectively without a year. But the day is ingrained into our biology. Even on the space station, where the sun rises every hour and a half, the crew still lives on a 24-hour clock. As was described in the last chapter, up until a few decades ago the day set the standard for time. But the Earth day is unique only in our tiny corner of the cosmos. If we lived on the surface of Jupiter our day would last only 10 hours from noon to noon. On Venus a solar day lasts about 117 Earth days, with the Sun rising in the west and setting in the east.

On Earth a day is 24 h, whereas the amount of time to rotate once around our axis when compared to the stars is 23 h, 56 min. The first is called a *solar day* and the second a *sidereal day* (see Figure 9.1). They are quite similar because our day is fast compared to a year. If you picture our orbit as a circle, after one rotation on our axis we have traveled about 1° around the circle, and so need to rotate an additional degree to see the Sun centered in the sky again.

The Moon is an interesting case. Its orbital time is equal to its rotation, which is different from the time between full moons. The orbital and rotation time is 27.3 days, the time from full moon to full moon,

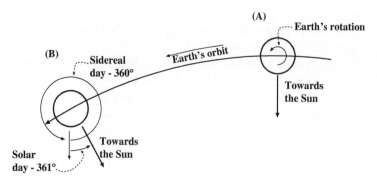

Figure 9.1 Solar versus sidereal time. Starting at (A) the Earth orbits the Sun and rotates on its axis to (B) a day later. After 23 h 56 min—a sidereal day—it has rotated 360° and faces the same stars. After 24 h—a solar day—it has rotated 361° and faces the Sun.

called the *synodic period*, is 29.5 days. The difference is like the difference between the solar and sidereal day, both caused because the Earth moves around the Sun. But what is so interesting about the Moon is the synchronization of the rotation and orbit, which we see because only one side ever faces the Earth. The Moon is not just a passing rock, accidentally in our orbit. Its formation and evolution must be intimately tied to that of the Earth.

Still, what I am looking for is a clock with which to measure long time periods. The time periods of the orbits of planets are another potential celestial clock. One significant difference between planet-days and planet-years is that there is a strong trend between the position of the planet and the length of its year. This is not true of the length of its day. The further out the planet's orbit, the longer its orbital time (see Table 9.2). At first we might think it takes longer because it has further to go. But Johannes Kepler (1571–1630) spotted the real trend here. As his third law of planetary motion says, "the square of the orbital period is proportional to the cube of the semi-major axis." This means that the outer planets not only have a longer distance to travel to orbit the Sun, but they are also traveling slower.

A good feature of using planet-years as opposed to days as a standard is that they are predictable: they are governed by Kepler's laws. If I am looking for a very slow celestial clock with which to measure the age of the universe, I will want to use something with a very large orbit.

Table 9.2 Days and years of the planets.

Planet	Sidereal day	Solar day	Year	Orbital radius $\times 10^6$ m
Mercury	58.65 dy	175.94 dy	88 dy	58
Venus	−243.02 dy	−116.75 dy	225 dy	108
Earth	23.93 h	24.00 h	1.00 yr	150
Mars	24.62 h	24.66 h	1.88 yr	228
Jupiter	9.92 h	9.92 h	11.86 yr	778
Saturn	10.57 h	10.57 h	29.46 yr	1,427
Uranus	−17.24 h	−17.24 h	84.33 yr	2,871
Neptune	16.11 h	16.11 h	164.79 yr	4,498
Pluto	−6.39 dy	−6.39 dy	246.04 yr	5,874

The negative sign indicates that rotation is in the opposite direction to the orbit. The most distant planets change their position relative to the Sun slowly, and so the difference between their solar and sidereal days vanishes. Times are in Earth-solar days and Earth years.

Neptune takes 165 Earth years to circumnavigate the Sun, and Pluto (even if it is not really a planet) takes 248 Earth-years to finish a lap. Archimedes wrote *The Sand Reckoner* only about 9 Pluto-years ago. But there are things in our own solar system that have even longer periods.

Comets have orbits just like planets, even if they are highly eccentric and elongated, but they are still governed by Kepler's laws. Most famous of comets is undoubtedly *Halley's comet*, named after Edmond Halley (1656–1742), Astronomer Royal of Great Britain. Halley realized that this comet returned every 76.7 years. Halley's comet shows up on the Bayeux tapestry, with ominous foreboding for King Harold and the Battle of Hastings (1066), about a dozen Halley's-comet-years ago.

However, Halley's comet barely ventures beyond the orbit of Neptune and so is a frequently seen comet. The comet *Hale–Bopp* has an orbit period of 2,400 years. That means that Stonehenge is only about two Hale–Bopp years old. The comet *Hyakutake*, which passed us in 1996, last passed us about 17,000 years ago. But it was caught by Jupiter's gravity and slingshotted into an even greater orbit. We should not expect to see it return for about 140,000 years. In fact, this event points out one of the problems with using comets as clocks. They can easily have their orbits altered by the gravity of the planets and other things they pass. Or they may cease to exist, like the comet *Shoemaker–Levy 9*, which collided with Jupiter in 1994.

Still, I am looking for a slower clock that could measure even more ancient time. So I need to look for something with even a larger orbit.

***

Nearly a billion times larger than the solar system are galaxies and within each one stars are orbiting the center of the galaxy. This should be the ultimate clock. Our Sun, in fact our whole solar system, is zipping around the Milky Way at 230 km/s. At that rate, and given the size of our galactic orbit, it will take about 230 million years to make one trip around.

A *galactic year* sounds like a good unit to measure out the eons of deep time. But like planetary days, it is not quite as universal as we would like. We sometimes think of galaxies as giant solid pin-wheels with stars embedded in them. But they are not solid and the stars are not rotating together. Since the stars are still in orbit we might still expect them to follow Kepler's laws. They do not follow the rule that "the square of the orbital period is proportional to the cube of the semi-major axis," but their motion is still governed by the laws of physics.

Kepler's laws occupy a unique niche in the history of science. They are very precise, and they are based upon observation. But when originally posed, they were without a theoretical foundation. Within that same century, Newton penned his gravitational law and derived Kepler's three laws of planetary motion. Deriving those laws in many ways vindicated Newton. You can derive Kepler's laws from Newton's gravity if you assume that the planets are orbiting a mass that is concentrated in the center of the system. In our solar system, over 99% of the system's mass is in the Sun. In galaxies, the mass is not so concentrated. You can still use Newton's law of universal gravitation, but you need to account for the distribution of matter, including the elusive dark matter. What we see is that orbital speeds are nearly constant for most stars in the galaxy, someplace between 200 and 250 km/s. Of course those stars out on the edge of the galaxy's spirals have a long path and so a long orbit time.

The solar system travels once around the Milky Way every 230 million years or so. We really are talking deep time now, with the age of the universe at about 60 galactic years. We do not know of any system larger than galaxies that have this sort of periodic, or almost periodic, motion, and therefore we will have to look for slower clocks elsewhere. Let us just quickly review our timekeepers up to now and see what we have missed.

At the fastest end of our list of clocks was the decay of exotic subatomic particles like the delta particle or the Z boson, which have lifetimes measured in yactoseconds, 10^{-24} s. Deltas form due to the rearrangement of quarks within a proton or neutron, and the transition times are driven by the strong force, the force that binds quarks into the particles that we can see.

A great deal slower than strong transitions are atomic transitions, measured in nanoseconds, 10^{-9} s. These are the transitions that give us light to see by and also the cesium-133 clock. These transitions are driven by the electromagnetic force. The electromagnetic force is several orders of magnitude weaker than the strong force, so it is not surprising that its timescale is also much slower.

Our last timekeepers are based on the orbits of celestial bodies. It is the force of gravity that holds these systems together. And much like the trend between the strong and electromagnetic forces, gravity is much weaker and the timepieces are much slower.

We have, of course, skipped the weak force. The weak force is what causes radioactive decays. We will look at all the forces in detail in the next few chapters, but for now it is enough to know that the weak force has a problem: it is *really* weak. It is also very short range and so operates in a world within a neutron or proton, a region overwhelmingly dominated by the strong force. However, it does have a role. The strong force can hold together three quarks to make a neutron: an up and two down quarks. What the weak force can do, for example, is push one of those down quarks to decay into an up quark. In this case a neutron has become a proton, one of the most basic weak decays.

The weak decay can also happen between electrons and their exotic, heavier siblings, the muon and the tau particle. The decay of the muon is often considered the prototypical weak decay. This is because the muon decay is not buried inside of a bigger particle, and so it is free of strong-force complications. The lifetime of the muon is about 2 μs (10^{-6} s), which fits in with our comparison to strong decays (10^{-24} s) and electromagnetic decays (10^{-9} s). It is a bit weaker than the electromagnetic force and much weaker than the strong force. But by itself that does not help us get to deep time.

When we look at the weak decay of the neutron to a proton we find a lifetime of nearly 14 min. This range represents a factor of a billion. But the weak decay exhibits even a broader range than that. At the fastest end of the scale is the tau particle, the electron's heaviest sibling, with a lifetime of 3×10^{-13} s. A decay in a chain reaction can take a shake, or 10^{-8} s. But I am looking for a slower clock, one with which I can measure the age of stars and the universe.

Rubidium-86 decays in 18.6 days and americium-241, the heart of many smoke detectors, has a half-life of 432 years. Perhaps the most famous radioactive element is uranium-238 with a half-life of 4.47 billion years. We are now talking about timescales similar to the age of the Earth. If you had a kilogram of U-238 when the Earth was formed, you would now have half a kilogram left.

One of the slowest decays measured is vanadium-50 with a half-life of 1.5×10^{17} years. It may not actually be the slowest decay; it is just hard to measure these rare events. Actually you do not have to start with a kilogram and wait for half of it to go away. Instead you start with a known amount and watch the rate of decay particles coming out. For example, if I had a milligram of vanadium-50 (which is hard to obtain) and I watched it for a year I would see about 56 decay particles radiating out. When I compare the 56 I saw to the number we started with (think Avogadro's number) we could work out the half-life. Measuring a long half-life comes down to isolating a good sample, and waiting for the occasional decays.

***

Perhaps the most famous use of radioactive decay related to time is carbon-14 dating. But before we can look at this we need to understand a little bit about carbon and isotopes.

Most carbon is carbon-12, often written as ^{12}C. In its nucleus are six protons and six neutrons. There are also six electrons orbiting the nucleus. It is those electrons that set carbon's chemical properties and that is how we recognize it as carbon. In fact all carbon atoms have six electrons and therefore six protons because they are electrically neutral. But the number of neutrons is not so well determined. In nature there are carbon atoms with six, seven and even eight neutrons. These different variations all act essentially the same chemically, which is why we call them all carbon, but they have different masses. We call these variants *isotopes*, a term which is based on the Greek words for "same place," meaning the same place on the periodic table of the elements. The 12 at the end of carbon-12 thus tells us the number of protons plus neutrons.

Carbon-13 has an extra neutron. It is like carbon-12: it is stable and will not decay and will participate in all the same chemical reactions as carbon-12. Carbon-14 has two extra neutrons that unbalance the nucleus a little and so it will eventually decay. A single carbon-14 atom has a 50% chance of decaying after 5730 years. That means that if I started out with a million atoms, after 5730 years I will have only 500,000 carbon-14 atoms left. After a further 5730 years I will be down to 250,000 atoms (see Figure 9.2). This is the basis for radiocarbon dating, which is so often used in archeology. If you know how many carbon-14 atoms an object started with, and you know how many it has now, you can figure out how long that bit of carbon has been around. The trick is to figure out how many were originally in the sample.

Carbon in a piece of wood comes from the atmosphere at the time the tree was growing. In the process of photosynthesis carbon dioxide

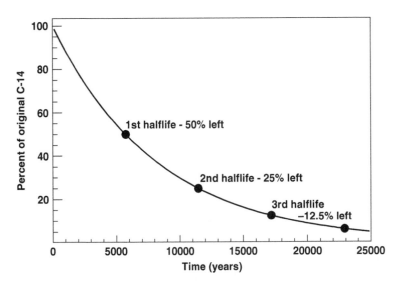

Figure 9.2 The decay of carbon-14 used for dating. Half of the carbon-14 in a sample decays after 5730 years, half of the remaining carbon decays over the next 5730 years and so forth.

(CO_2) was absorbed by the tree's leaves and was synthesized into sugar and then into cellulose. If all carbon-14 on Earth dated from the time of the origin of our planet, it would all have essentially decayed away by now. But carbon-14 is constantly being produced in the upper atmosphere. When a high-energy cosmic ray collides with a nitrogen atom it might transform that atom into carbon-14. This means that there are always new carbon-14 atoms being created and the number in the atmosphere is nearly constant. Our atmosphere has one carbon-14 for every trillion carbon-12 atoms. So by measuring the ratio $^{14}C:^{12}C$ in your sample now, and knowing that it was originally one to a trillion, you can figure out how old the sample is.

This whole technique seems almost too clever. It is also true that people have been tweaking it since Willard Libby developed it in 1949. For instance, if the object is an important artifact, such as the Shroud of Turin, you will want to use as little of a sample as you can. Also, how do we know that the production of ^{14}C has been constant? We can actually confirm this number by looking at the carbon-14 content of objects with a known history. Libby was able to get a scrap of wood from an ancient Egyptian barge, the age of which was known from written records. He could also confirm the technique by looking at the heartwood of a redwood tree that had been cut in 1874. By counting tree rings he could independently determine the age of the heart wood as 2900 years old, in agreement with the carbon-14 dating results.

Actually, by looking at tree rings people have found slight variations of the trillion-to-one number over time and in various locations, but nothing dramatic. With a lifetime of 5730 years, carbon-14 is well suited to dating objects through the history of modern man. In fact we can date samples back to about 63,000 years, but not much beyond.

***

We now enter realm of geologic time, the time it takes to build mountains, cut valleys, or even form the Earth itself.

Once you have watched a flood carve out a new channel, or a volcano erupt and new rocks appear it soon becomes apparent that the world is in flux. When you look at the side of a mountain, especially in a dry region or if it is very steep and so not hidden beneath trees, there is a story to read. Layers of sandstone, limestone, or schist must be telling us something, if only we knew how to read it.

One of the first people to try and read these layers was Nicholas Steno (1638–1686), a Danish naturalist and later a priest and bishop. He recognized some of the most basic principles of *stratigraphy*, principles like: strata are laid down horizontally, even if later they are tilted. Also, newer layers must be formed on top of older layers. So when you see a deep cut through the Earth, like the Grand Canyon (see Figure 9.3), the newest rocks are on top, near the canyon's rim, and the oldest are

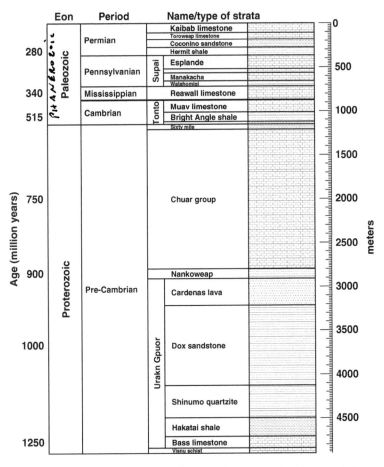

Figure 9.3 The layers or *strata* of the Grand Canyon. Fossils in each layer order and date the rock. Radiometric dating is also used.

at the bottom. The rimrock is Kaibab limestone, which is about 250 million years old. At the bottom of the canyon, near the Colorado River, is the Vishnu schist, which is almost two billion years old.

Strata are not always simple to read. Over time a layer may be eroded before a new layer caps it, such that a layer might be thinner than it was originally, or even missing. As continents collide and mountain ranges push up layers can get twisted, tilted, or even inverted. But the general trend is old on bottom and new on top, a feature which geologists use to decipher the history of the Earth.

In the century following Steno, people were collecting rocks and also fossils, but it would be a mining and canal engineer in England who really saw the important relationship between strata and fossils. William "Strata" Smith (1769–1839) realized that the same fossils show up in the same strata. He also realized that the same strata existed all across the British Isles, but sometimes tipped, so that what was near the surface in Wales was buried deep under London. This meant that once the fossil record was established where rocks were undisturbed, then rocks that were broken, tilted, or even inverted elsewhere could be dated by their fossils, at least in a relative sense.

When you start to read the fossil record as a book the story has a curious structure. The fossils change slowly as you move through a stratum, but then there are jumps. A species might disappear, or sometimes a whole group of species became extinct. These chapter breaks are distinct and are seen across the world. Slowly geologists pieced together the pages, sections, chapters and volumes of geologic history. The time units that were established were the *period*, which was greater than the *era*, and which was in turn greater than the *eon* (see Table 9.3).

Our present eon is the *phanerozoic* (from the Greek for "visible life") and for a long time it was thought that this time period spanned the whole of the fossil record, from 540 million years ago to the present. There are no hard-shelled, macroscopic fossils before it. The oldest

Table 9.3 Present geologic time.

Eon	Phanerozoic	540 million years
Era	Cenozoic	65 million years
Period	Quaternary	2.6 million years
Epoch	Holocene	12,000 years

period in the Phanerozoic eon is called the *Cambrian* period (541–485 million years ago). Cambria is the Roman name for Wales, at the west end of William Smith's survey. It was also about the end of the hard-shelled fossil record. So the time before the Phanerozoic eon is called *Precambrian* eon. By the end of the Precambrian super-eon the world was already old. The Earth was covered with bacteria and some multicellular life, but these do not make the sort of fossils rock hounds would like to hunt.

Using stratification and the fossil record, geologists and paleontologists could name and identify the various times in geological history. They could also order the eras. They could even make estimates of lengths of time by noting how fast a modern lake or sea will acquire silt or sand and then extrapolating that knowledge to the thicknesses of shales and sandstones. But these were crude estimates, subject to numerous unknown factors and assumptions.

Ideally, one would like to use a technique like radiocarbon dating to establish the age of a rock. The difficulty is that you need to know how much of an isotope there was in a rock initially. You also need an isotope with a very long lifetime, and there are many that meet that criterion. Actually, several techniques have been devised but I will only describe one based on the potassium and argon levels in a rock.

Argon is an inert gas that would not bind to a rock if it was fluid, so a new igneous rock will be argon-free. However the potassium-40 in the rock can decay into calcium and argon and after the rock has solidified that argon will be trapped within the rock. So a geologist can break open a rock and measure the ratio of potassium-40 to argon-40. By knowing the lifetime of potassium, and what percentage of it decays into argon and to calcium, the age of the rock, or at least the time since solidification can be calculated. Since the half-life of potassium is 1.25 billion years, this technique will work across a wide range of times, but with less precision than carbon-14. Still, it is well matched to geological eons/eras/periods.

So now we can put absolute dates on geological time. The great dinosaur extinction shows up as the border between the Cenozoic and Mesozoic eras, about 65 million years ago. The start of the hard-shelled fossils, the *Cambrian revolution* is about 540 million years ago. The oldest microscopic fossils are about 3.5–4 billion years old.

***

The life of the Earth is not just about life on the Earth. The Earth itself evolves and changes shape. Continents have merged into super-continents and then split apart. Oceans have swelled and vanished. It is not that hard to believe that the eastern most point of South America, Cabo de Sao Roque in Brazil, would fit into the Gulf of Guinea in Africa, but when Alfred Wegener first suggested continental drift in 1912 it was seen as radical. However, we now realize that it is not just that the coastlines seem to match, but also that the mid-Atlantic ridge is made of young rocks, and we have actually measured the spreading of the Atlantic. We can measure the motion of all the continents and extrap-olate back 200–300 million years ago when they were joined together into one supercontinent called *Pangaea* (see Figure 9.4).

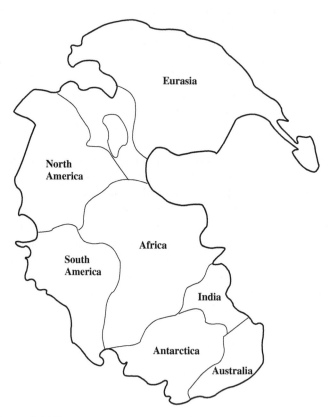

Figure 9.4 The supercontinent *Pangaea* about 200–300 million years ago.

The Earth seems so solid and rocks seem so rigid and permanent. But given enough time—and we are talking about hundreds of millions of years—and the heat that rises up from the core of the Earth, the continents are plastic and malleable. We should think of the Earth as being in a slow—very slow—boil, with rocks rising in one place and being drawn down in another. Giant convection currents exist, only with continental plates instead of air or water.

Pangaea was not the first and only supercontinent. Earlier in geologic time there were Rodinia, Columbia, Kenorland and Vaalbara as well as a number of other continents and fragments (see Table 9.4). Throughout the history of the Earth the continents bump into each other every few hundred million years, stick together for a while, and then move apart.

The further back in time we go the harder it is to piece together the evidence about the continents, but we can still date rocks. Some of the oldest rocks found are from Australia and Canada. In Canada, the *Acasta gneiss*, an outcropping in the Northwest Territories has been dated to 4 billion years old. Also the *Nuvvuagittuq greenstone*, from near Hudson Bay may be older, but having more complex chemistry, its radiometric date is still being studied. Older than these two are small crystals of Zircon found in the Jack Hills of Western Australia, which date back 4.4 billion years.

There are in fact older rocks on Earth, but not of Earthly origins. Famous among these is the *Genesis rock*, which was brought back from the Moon by the Apollo 15 mission. It has been dated at 4.46 billion years. And even older than that are meteorites. We think that meteors were formed at about the same time as the Earth and solar system and have remained unchanged since then. While the Earth has continued to roil, these meteorites have remained intact for 4.54 billion years. In

Table 9.4 Supercontinents through the history of the Earth.

Supercontinent	Millions of years ago
Vaalbara	3,100–2,800
Kenorland	2,700–2,500
Columbia/Nuna	1,900–1,400
Rodinia	1,100–750
Pangaea	300–200

The list is incomplete.

fact 4.54 billion years (or 1.43×10^{17} s) is our best measurement of the age of both the solar system and the Earth, to within about 1%.

***

We have walked back about a third of the way to the Big Bang. The universe is about three times older than the solar system. This comes as no surprise when you realize that at a time shortly after the Big Bang the universe was essentially made of only hydrogen. Yet here on Earth we have heavier elements; we have lead and gold and uranium.

A few stars can actually be dated with radiometric techniques. This is difficult since the isotopes generally appear as faint lines within the stars' spectra, especially compared to more common and stable isotopes. Still, the ratio of ^{232}Th:^{238}U has been measured in a few stars, yielding ages of over 13 billion years.

Radiometric techniques only work because we can measure the lifetimes of these isotopes here on Earth. Also the more we have learned about nuclear physics in laboratories the better we are able to construct models of how fusion works within stars, how fast elements are formed and what happens in a supernova. Our present models of the universe start out with the Big Bang. A few hundred million years later the universe was made primarily of hydrogen and a bit of helium. Stars were formed and started to burn that hydrogen. Through nuclear fusion those hydrogen atoms combined into helium and released energy, in the same way our Sun works today. This process of creating heavier elements is called nucleosynthesis: helium burns to form beryllium, and then carbon, nitrogen, oxygen and other elements. But the process of fusion can only go so far up the periodic table and it stops at iron. You cannot fuse iron with another atom and get energy. Iron is the most stable element. Yet we have heavier elements around us.

The only way to produce a heavier element is to put energy into the system, something which can only take place in a supernova. So since our Earth contains elements heavier than iron, we must assume that we are made of the ash of a previous generation of stars. Or as Carl Sagan was famous for saying, "We are made of star stuff."

This actually makes that radiometric dating of those old stars interesting. They could not have been first generation since they contained very heavy elements.

We can also measure the age of a globular cluster of stars by looking at the luminosity of the brightest stars. We can do this because we understand the way burn rates, mass, age and brightness are related to each other. By this technique the oldest globular clusters have been dated at 11.5 ± 1.3 billion years.

What has been encouraging about this whole field is that independent measurements are converging upon similar estimates of the age of the universe. It is true that these results are dependent on models of stars and nucleosynthesis, but those models are becoming more robust each year. They also calculate more than star ages. They predict the relative abundance of all of the elements and isotopes in the universe, in good agreement with what is observed.

One last technique for establishing the age of the universe is to look at the rate of expansion. As we will see in a later chapter, we can measure the distances to galaxies and the rate at which they are moving and extrapolate back to the moment of the Big Bang.

All of these techniques: radiometric dating of stars, luminosity of stars in globular clusters, nucleosynthesis and the abundance of elements and finally the expansion of the universe all agree 13.7 billion years as the age of the universe. I actually find it amazing that the Earth has been here for a lot of that time, about a third of the life of the cosmos.

***

This is a good time to pause and look at where we have been in these last two chapters. At one end of our time scale are the lifetimes of exotic, subnuclear particles such as the delta particle or the Z boson. Their lifetimes are measured in yactoseconds, 10^{-24} s. This is about the time it takes for light to cross a proton. The slowest thing we have seen, the event that has taken the longest time, is the evolution of the universe itself. The age of the universe is 13.7 billion years or 10^{17} s. It takes light about that long, to within an order of magnitude, to cross the universe that we can see. It is not a simple coincidence that time, size and the speed of light are connected like this. For something to evolve, no matter what its size is, its parts have got to interact with each other, and no interaction is faster than light.

In the next four chapters we will look at the size of an array of objects and structures, much as in Chapter 2 we looked at the size of biological organisms, and we will try to find some organizing principle. We will try to be aware of the way things interact and forces cause structure. The world is not just a haphazard jumble of objects. There is order; we just need to see it.

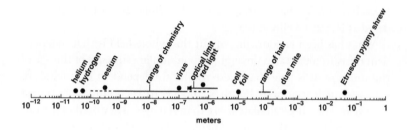

❧ 10 ❧

Down to Atoms

Poems can have meters and one may judge an event with a metric, but neither of these has much to do with a platinum–iridium bar in Paris. One can meter out punishments or rewards or rations or even electricity. We can learn to keep the beat or meter of a song with the aid of a metronome. Yet none of these has anything to do with a hundred-thousandths of the distance between the Earth's equator and pole.

The word meter (or metre) and metric are both based on the Greek word *metron*, to measure. We often think of the metric system as a system based on multiples of ten or a thousand or a million, but the word itself only tells us that it is a system for measurements. Its official name is *Le Système International d'Unités* or, in English, *The International System of Units*, often shortened to just SI. The system could have been many things. But its most important feature is that it is recognized internationally. It could have been based upon the mile, the hour, or the calorie. The basic length unit could have been set by the height of a student at MIT named Smoot. In fact the bridge next to MIT that connects Cambridge and Boston has been measured as "364.4 Smoots, plus or minus 1 ear." But that unit is hard to reproduce.

Today the meter is defined as the distance over which light can travel in 1/299,792,458 of a second, and a second is defined as the amount of time it takes for 9,192,631,770 oscillations of a Cs-133 atom. This may seem like a convoluted way of defining our most basic unit of measurement. But it meets the original criteria that the enlightened thinkers of the French revolution recognized as really important. A meter could be created by anyone anywhere. If we were in radio communications with aliens from another galaxy we could describe our meter and they could produce a meter stick of exactly the right length. But a combination of light and cesium clocks is not how the meter started.

***

Back in Chapter 1 we briefly introduced the meter and its origins. It was designed to be 1/10,000,000 of the distance from the equator to the pole. It was an inspirational goal, a majestic plan. It embodied some of the ideals of the Enlightenment, it was universal and it belonged to all mankind. So Delambre and François measured part of a meridian in France and the meter was established. On December 10, 1799 that meter was inscribed upon a platinum bar that was deposited in the National Archives of France.

The meter was inspired by the 1/10,000,000 of a quadrant, but the real meter was a metal bar. Within a few years the accuracy of the Delambre and François survey was in doubt, but the bar remained.

By the middle of the nineteenth century the need for an international measurement standard became more and more apparent as machinery and trade increased. So in the 1870s France hosted a series of meetings and conferences that established the meter as the international standard. Even the US was a signer of that original treaty. It is true that the inch, foot and mile still dominate in the US, but did you ever notice that an inch is *exactly* 2.54 centimeters. In fact the inch is defined in terms of the meter.

At the September 1889 meeting of the General Conference on Weights and Measurement (CGPM) a new meter was established. This one was made of a platinum–iridium alloy. The original meter bar was showing its age. Its ends had signs of wear after years of being handled when copies were made to be sent out across the globe. At this time the methods of using the standard bar were also more tightly defined. The bar was only a meter when it was at $0°C$, at the freezing point of water.

By 1927 there were a few additional caveats for an acceptable measurement. A standard air pressure was established and it was specified that the bar must be supported in a prescribed manner. These refinements may not seem very important, but what they tell us is that there was a rising ability to make precise measurements, as well as a need for greater accuracy. Yet still the meter was based on a metal bar in Paris.

The standard changed radically on 20 October 1960. The CGPM redefined the meter as 1,650,763.73 wavelengths of light in a vacuum from the $2p^{10}$ to $5d^5$ transition of krypton-86 atoms, a jump between two atomic orbits. This is a nine-digit number, which means that the meter was defined to within about ten atoms. But more importantly, the meter no longer resides solely in Paris. The meter had been freed from that metal bar. Anyone, anywhere could make an identical meter stick.

That, however, is not the end of the story. On 21 October 1983, the meter was abolished. The meter is no longer considered a primary unit of measurement. A meter is now derived from the second and the speed of light. In 1983 distance could be measured to 1 part in a billion (10^9) whereas clocks were approaching one part in one hundred trillion (10^{14}). In fact clocks are now a hundred times better than they were then.

With the speed of light defined as 299,792,458 m/s exactly, it takes light exactly 1/299,792,458 seconds to travel 1 m. Ideally the meter would now be as accurate as clocks: one part in 10^{16}. In reality it may not make sense to think about the length of an object in units smaller than atoms.

Actually in 2002 there was one more caveat added to the definition of the meter. The meter is a *proper length*. Proper length is a concept that comes out of relativity and the 2002 convention says that a meter is only well defined over short distances where the effects of general relativity are negligible. In other words, if you are measuring an object next to a black hole, or a length that spans a galaxy, you should not expect many meaningful digits of accuracy.

The standard meter bars produced in the 1880s are really objects of beauty: metallic sculptures. The bars are a bit over a meter long with marks inscribed upon their sides. By modern measurements those marks are placed with a precision of 10^{-7} meters. The bars are made of platinum–iridium, a silvery colored metal, and its cross section is a modified X, a shape now called a Tresca, after Henri Tresca, who suggested its use because of its rigidity (see Figure 10.1). These days such bars are museum pieces, retired since 1960. They are no longer doing what a meter should be doing. Meters are meant to measure things by their nature and name.

Figure 10.1 The cross-section of a standard meter bar from the 1880s. The design was proposed by Henri Tresca and is very stable.

Before we can measure things, especially small things, we will need to subdivide our meter into smaller units. We can use simple geometry to subdivide our meter stick into ten subdivisions called *decimeters*. Geometry will also allow us to divide decimeters into ten centimeters (10^{-2} m) and each centimeter into ten millimeters (10^{-3} m). A centimeter is about the size of a periwinkle, the Etruscan pygmy shrew, or the bumblebee humming bird, which we met in Chapter 2. It is also the minimum size of a warm-blooded, or homeothermic, animal.

Equipped with a meter stick marked in centimeters and millimeters we can now measure some amazingly small things, including the thickness of paper, foil and hair. Once we know the thickness of a hair, we can then measure the wavelength of light and the size of oil molecules, an amazing accomplishment for a macroscopic stick.

***

We start with paper. If you turn up the edge of the piece of paper you are reading this on and hold it next to a meter stick all you can really see is that it is much smaller than that smallest division; it is smaller than a millimeter. But we can do better if we are a bit cleverer. Close the book for a moment and measure the thickness of all the pages, not including the cover. Now all you need to do is take that measurement and divide it by the number of pieces of paper in the book. An easy way of finding that number is to take the last page number and divide by two because pages are printed on both sides. The first draft of this book was written in a notebook 1.5 cm thick with 150 sheets of paper, so each sheet of paper is 10^{-4} m thick; that is, a tenth of a millimeter or 100 μm.

A bit more of a challenge is the thickness of cooking foil. The outside of the box claims that it contains 30 m^2, and my meter sticks tells me it is 30 cm wide, so the roll of foil must be 100 m long. If I look at the end of the roll I can estimate that each wrap is about 10 cm long, and so there must be about 1000 wraps. My meter stick can also measure the thickness of the foil on the roll at about a centimeter. So if 1 cm is 1000 layers of foil, then 1 layer is about 10^{-5} m. This is also the diameter of an average cell.

We have just measured something the size of a cell with a meter stick marked to only millimeters, and that measurement is really pretty good. Of course the reason for our success is that we actually measured a large collection of paper or foil and then did a bit of division. This is a technique we will continue to use for smaller and smaller objects.

If we could stack hair like we can paper or foil we could use the same technique to measure a hair. But hair tends not to stack so simply. However, there is a technique that we can use and which is often attributed to Isaac Newton. Start with a long hair—from portraits I have seen of Newton his was very long. Now wrap that hair in a nice neat coil around a pencil, counting the number of turns as you go. Make sure each winding is snug up against the previous winding. If you have dark brown, almost black hair you will find that you have about ten windings for every millimeter of pencil. The more windings you make, the better your measurement. Ten windings per millimeter means the hair is 10^{-4} m thick, or 100 μm. As we saw in Chapter 2, hair ranges from about 50 μm for light blond hair to 150 μm for thick black hair.

***

Now we would like to measure the size of a light wave, which is really small. We can measure it relative to a hair and, since we know the size of a hair in meters, we will know the size of the light wave. We can do this because light is a wave and follows the same rules as sound or water waves. Waves have two important properties for our measurement; they spread out around corners, and when two sets of waves meet they add up or *interfere* with each other.

Sound clearly spreads around edges and corners. That is the reason you can hear something you cannot see, especially outside where there are few walls for sound reflection. Also if you watch an ocean wave pass a jetty or breakwater, it too spreads out and can roll towards the corners of the harbor. We do not notice light doing this, but it does. One of the big differences between light waves and sound or water waves is their size. Water waves can measure tens of meters from peak to peak. Sound waves range from a meter to a millimeter. Light waves are about 10^{-7} m, a tenth of a micrometer or a thousandth of a hair's breadth. In order for a wave to wrap around a corner, that corner must be sharp compared to the size of the wave. So a razor blade, or a hair can deflect light that passes by it.

The other wave phenomenon that we need is interference. When two waves meet each other they can combine *constructively* or *destructively*. Constructive means that the peaks line up in the same place, at the same time, and add together to form a bigger wave. Destructive means that the peak of one wave lines up with the trough of the other wave and cancel each other. This phenomenon can be dramatic with sound waves. If you have two speakers set a meter or so apart and playing a monotone, you can move your head around and hear places where the sound is stronger or weaker. The distance between louder, constructive nodes and quieter, destructive nodes is related to the wavelength. Real music is made of all sorts of waves and so the nodes for every frequency are scrambled all over the place so you generally do not notice interference in normal circumstances.

Water waves will of course do the same thing. If a wave passes an island or a large rock the waves will wrap around the island and meet each other at the back. There are places where the waves add and other places where they cancel. The way they join is called an *interference pattern* and its shape depends upon the size of the island and the wavelength of the ocean waves.

Light does the same thing. White light, like music, has a lot of frequencies. But if you take a monochromatic light, like a laser pointer, and shine it on a single hair, most of the light will go by the hair unaffected. A small part of it will be bent by the hair, and then interfere with the light going around the other side of the hair. On the wall you will see a strong spot where most of the light went, but also a series of spots. The spacing of that series depends upon the color or wavelength of the light and the thickness of the hair.

Interference is such an importance phenomenon that I will describe it with one more analogy. If a precision marching band goes around a street corner it generally performs something called a *wheel* (see Figure 10.2). In order for the rows of marchers to stay in line the band members near the corner slow down. Now let us imagine that the band members can only march at one speed, for our band is part of a light

(A)   **(B)**

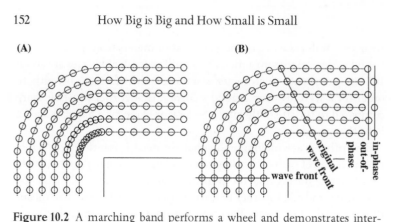

Figure 10.2 A marching band performs a wheel and demonstrates interference. (A) A marching band performs a wheel. (B) A marching band with constant speed. After the corner, columns 1 and 3 are in phase and constructive. Columns 1 and 2 are out of phase and destructive.

beam and light has only one speed. After the turn the rows will be messed up because the outside marcher went farther. However it is possible that marchers in the outside columns might line up with members of a different row if the extra distance they went was a wavelength, or a row gap. Actually if the extra distance is one, or two, or three, or more row gaps they will line up. But if it is a half, or one and a half, they will not.

Back to our light and hair. If we shine a laser on a single hair, most of the light passes by the hair and forms a bright spot on a distant wall. But some of the light will be deflected by the hair (see Figure 10.3). When photons from the right-hand side of the hair and photons from the left-hand side reach the wall they can form an interference pattern. The first

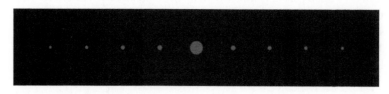

Figure 10.3 Laser light on a single hair shows an interference pattern. When a red laser is shone on a single hair most of the light goes straight (center spot), but some of it will form an interference pattern, which is seen as spots next to the center one. If it is about 10 m from the hair to the wall, the spots are spread out by 5–10 cm, depending upon the hair and laser.

spot on the side of the major spot is where light from one side of the hair traveled exactly one wavelength farther than light from the other side. The second spot means that the light's path length was different by two wavelengths. That means it is just a bit of geometry to measure the wavelength of this light. We find that red light has a wavelength of about 6.7×10^{-7} m, blue light about 4.7×10^{-7} m.

***

From the meter stick we measured the size of a hair, from a hair we can measure the wavelength of light. Now, knowing the size of light, we can measure the thickness of some molecules.

We have all seen puddles on roads with a very thin film of oil. Sometimes we view them as unsightly pollution, other times as beautiful swirls of color. What is happening is that sunlight is shining on that film and some of that light is reflecting off the surface and towards the eye. The rest of the light passes through the oil film to the surface of the water where again some of it is reflected up towards the eye. Light off the oil's surface and light off the water's surface travel slightly different distances. If the distance is a wavelength the light combines constructively and we see it brightly.

When you look at different parts of an oily puddle you are looking at different angles and also at different thicknesses. The angle and the thickness will select the wavelength and so the color of light you see. Measure the color and angles and you can figure out the thickness of the oil. For some oils that thickness may be the size of a single molecule. Oil is typically $1-2 \times 10^{-9}$ m thick.

Actually Benjamin Franklin first observed the way oil spreads on water and Lord Rayleigh measured the thickness by seeing how far a single drop of oil would spread. Since he could measure his oil drop as a sphere and the surface of a flat oil slick with a meter stick, Lord Rayleigh's method gives us the size of the molecule directly. But en route we have touched on interference, something we will continue to use.

***

The world smaller than cells yet larger than atoms is the world of chemistry. This is a complex place where atoms can combine in a plethora of ways to build innumerable types of molecules (although actually, since the number of atoms in the observed universe is finite, the number of combinations is also finite and even calculable.) So I will not discuss particular molecules like benzene rings or carbon chains. What I will talk about is why they stay together and how that affects their shape and size.

Why is it that molecules do not just fall apart? Atoms are bound together into molecules by chemical forces and chemical forces are derived from electromagnetic forces. Most atoms are electrically neutral yet they still bind, a problem alluded to in Chapter 7.

The most important point of electromagnetism is something we all learned with bar magnets in elementary school: opposites attract and likes repel and neutrals do neither. So when an electron passes by an atom, it sees the atom as neutral and so it is neither attracted nor repelled. In fact, it is attracted to the atom's proton and repelled by its electron, effects which tend to balance out. However, if it is a close encounter things change. The electron's trajectory could take it closer to the atom's proton than the atom's electron and so it would be slightly more attracted. It might even push on the atom's electrons, distorting their orbit. Think of the atom's electrons in a dance around the proton. An electron from the neighboring atom might be close enough to 'cut-in' and join the dance.

When two or more atoms are near each other we can picture them as a group of dancers. Dance partners are primarily attracted to their partner but they are not blind to the opposite sex in other dancing pairs. Are they only attracted, or do they sometime cut-in and switch partners? When nature has a choice she usually chooses both. Sometimes atoms are mildly attracted from a distance, sometimes partners are shared, and sometimes dance partners are stolen outright, leaving the abandoned partner orbiting the central waltz. All of chemistry is just understanding the attraction, repulsion and dynamics of a dance floor. The different solutions—shared, borrowed, or stolen partners—are analogous to the various types of chemical bonds. What this tells us about size is that chemical bonds are short, no more than the size of an atom. If the atoms were well separated, they would look neutral to each other and not interact. If they were too close, the atoms would get mixed together. So bond length should be about an angstrom, 10^{-10} m.

As an aside, the angstrom is a measurement unit often used in atomic physics and chemistry. It is defined as 10^{-10} meters. It has the symbol Å and was named after a Swedish physicist Anders Ångström who was studying the spectrum of the Sun. The symbol, Å is part of the Swedish alphabet and, in English, is called an *A-ring*. The angstrom is close to the size of atoms and bond lengths. The diameter of a hydrogen atom is about 1.12 Å and if two hydrogen atoms are bound together they are separated by 0.74 Å. Most chemical bonds are between 1 and 3 Å.

One of the consequences of short bond lengths is that atoms tend to bind only to their neighbors. If two carbon atoms in a benzene ring or a hydrocarbon chain have another carbon atom between them, they do not really affect each other, except in that they may modify that intermediary atom. Chemistry is essentially about atoms binding to neighbors and so the models people build of molecules out of balls and rods are pretty good pictures. Molecules have structure. They are not just amorphous collections of atoms.

To picture a large molecule we can again return to our dance analogy. Imagine a large room full of square dancers, where each square represents an atom with a number of electrons. In this dance members of one square will occasionally switch with members of an adjacent square. This means that squares (atoms) need to stay near each other in order for dancers (electrons) to be shared or swapped in a short time. The molecule can be quite complex, with squares of different size representing different elements. An individual electron dancer may migrate across the whole floor, but the squares need to stay put for the dancers to find their place; in other words molecules are stable.

***

The smallest molecule will have the least number of atoms (two), the smallest atoms (hydrogen or helium) and the shortest bond length. Since helium does not chemically bind, the smallest molecule is H_2.

The largest molecule is essentially without bounds. An atom can be bound to two other atoms and form a chain, or more than two and create a complex web-like structure. There are molecules with millions of atoms. A single strand of DNA can have several billion atoms (10^{10}), but only occupy a space about 1000 Å across. This brings us back to the dimensions of a virus, a measurement that we looked at in Chapter 2, which is not surprising since a virus is primarily a strand of DNA (or RNA).

There are molecules even bigger than those found in biology. Plastics can be made from long, chain-like molecules and carbon nanotubes and fibers are in some sense mega-molecules. Larger than these are crystals, which are atoms held together with chemical bonds. Actually people are hesitant to calls these true molecules, because they have repeating patterns without limit and they do not really follow the law of definite proportions, which Dalton (Chapter 2) saw as critical. Still, they are very interesting.

***

In the world of mathematics crystals are cool: they have lots of symmetries. This is because crystals are made up of a grid of atoms, each atom linked to its neighbor in a regular pattern that is repeated over and over again, much like tiles on a floor. If the tiles are square or hexagons or triangles, that will dictate the pattern of the whole floor. If the first few atoms are arranged at right angles to each other, in a cubic structure, the macroscopic crystal will also be made with right angles. NaCl, common salt, is a go example of this. If the original was a tetrahedron, that will also show up in the macroscopic crystal. When you look at a quartz crystal you can see the tetrahedral angles, the angles between the bonds of silicon and oxygen.

Perhaps the most classic crystal is a diamond. As hard as hard can be, diamonds are the definition of 10 on the Mohs scale. They are made up of carbon atoms and carbon can bind to four adjacent atoms. When you try to link up carbon in a regular, three-dimensional grid you find that all the angles are the angles that show up in an octahedron. An octahedron is a three-dimensional shape with eight faces and each face is a triangle. It looks like two Egyptian pyramids with their bases joined together. This regular shape, starting at the atomic scale, is repeated again and again, trillions of times, until the crystal is big enough to see. Put together 10^{22} carbon atoms and you have a one carat diamond. This octahedral shape is the natural shape of a diamond.

A diamond like the Great Star of Africa, among the Crown Jewels in the Tower of London, or the Hope Diamond in the Smithsonian Institute are not octahedrons. They are cut diamonds. In fact the classic shape of a diamond that we associate with engagement rings is a *round brilliant cut*. This is one of many ways of cutting a diamond. The way you cut a diamond is actually dictated by the underlying atomic structure. There are preferred planes, or *facets*, along which you can cleave the crystal.

One way to think of a facet is to picture an orchard where the trees have been planted in perfect columns and rows (see Figure 10.4). Here

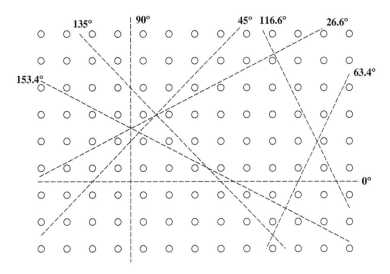

Figure 10.4 Trees in an orchard demonstrate the facets in a crystal. Where the trees are in perfect columns and rows there are special angles where you can see all the way through the orchard. These sight lines are analogous to facets in a crystal.

a facet is where I could draw a long line through the orchard and not encounter a tree trunk. If I look between two rows I can see to the other side of the orchard, so that is a facet. I can also sight between columns. I can also look at 45°, on the diagonal, and see a facet. This facet is like the way a bishop moves on a chess board. There is even a facet at about 26°, what you might call the knight's facet, over two columns and up a row. But you cannot just look at any angle. Trees do block the view in some directions.

Crystal facets can be used to cut gems, and the angles of the cuts can tell us about the underlying atomic grid. Actually, the facets of a diamond are more complex than that of an orchard, because the atomic grids are three-dimensional, but the reasoning is the same.

★★★

One last curiosity of crystals is that their regular pattern of planes and facets can also interact with light to form interference patterns, much like our examples of lasers and hair, or sunlight on oily water.

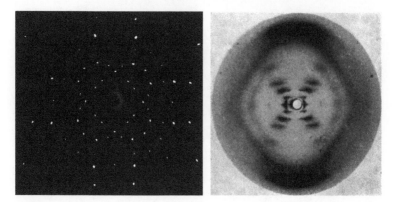

Figure 10.5 The interference pattern of X-rays from a crystal and DNA. (Left) silicon crystal, courtesy of C.C. Jones, Union College, Schenectady, NY. (Right) DNA, an image created by Rosalind Franklin and R. G. Gosling and which represents some of the first evidence of the double-helix structure. Reprinted by permission of Macmillan Publishers Ltd, *Nature* April 25, 1953.

We cannot use visible light because the light waves are much to big compared to the distance between facets. But if we shine X-rays with a wavelength similar to the atomic spacing we can produce beautiful interference patterns (see Figure 10.5). These patterns remind me of snowflakes, a kaleidoscope of bright spots in all directions, which have encoded in them the facet angles and spacing within the crystal. In fact it was using X-rays on DNA and looking at the resulting interference patterns that first pointed the way towards their double-helix structure.

***

Our first vision of an atom is of a nucleus in the middle, made up of protons and neutrons and electrons zipping around. Let us correct that image right away. The nucleus is about 100,000 times smaller than an atom. If you drew an atom on a computer screen the nucleus would occupy a fraction of a pixel. Seeing the nucleus inside an atom is like looking for a single sheep in the center of Wales. Our image really should be electrons zipping around a minuscule point in the middle. The size and shape of the atom is the size and shape of the electron orbits.

What we know about orbits is based on our experience with planets and satellites. If you toss a satellite into space with enough speed or energy it will settle down into a nice orbit. With too little energy it will come back to Earth, and with a bit more energy a wider, larger orbit. However, if you give it too much energy, a speed that is greater than the *escape velocity* you will lose your satellite, or perhaps you have just launched a deep-space probe. So orbits happen when objects have the right speed or kinetic energy. The Moon, or other satellites around the Earth not only curve through space, but they come back to their starting point, repeatedly. This is the vision Niels Bohr (1885–1962) first gave us for the structure of the atom: electrons in nice ring-like orbits about the nucleus; a planetary system. But this is not the way nature acts at this scale. At the size of the atom and smaller the world is governed by the rules of quantum mechanics.

Quantum mechanics has two unique features that are required to describe nature at these scales. The first is embedded in the name of this theory. The world is *quantized*, which means that it happens at certain values and not other values. When Bohr drew his atom he took this to mean that for some unknown reason electrons could orbit with certain allowed energies and with certain well-defined circular trajectories. Also it meant that there were energies and orbits where electrons were not allowed. This is different than for a satellite. If you add a little energy to a satellite it will go slightly faster and occupy an orbit that is slightly larger. The amount of energy you could add is continuous, not quantized.

The second unique feature of quantum mechanics is that matter, including electrons, has a wave-like nature. In fact this particle–wave nature explains quantization and a lot more. But before we look at quantum waves I want to step back and look at a guitar string and the wave it produces.

***

The lowest note on a standard guitar is made when plucking the sixth string, the thickest string, usually on the bottom. If unfretted it will produce an E_2 note, which means that it vibrates at 82.41 Hz. The string is of the right material and tension such that after you pluck it, it will produce that note for a while. That is not the frequency with which you plucked it. In fact, plucking or strumming a guitar is a violent act. At the point where it was plucked, the string is vibrating with all sorts of frequencies. Consider one frequency and the wave associated with it. It will travel up to the nut and down to the saddle, where the waves will bounce and propagate back to the middle of the string. If the frequency is on tune, when the two waves meet they will support each other, add constructively and continue to propagate up and down the string. If they do not match, the waves will be out of phase (one on the rise and one falling) and they will cancel each other and that frequency will quickly die out. When you tune a guitar you are changing the tension, which changes the propagation speed, and so selects a certain frequency. Electron orbits in atoms are selected the same way.

Electrons have waves associated with them, so when we picture an orbit we should think of a wave flowing around the nucleus of an atom. If the electron has the wrong wavelength when it wraps around the atom it will not be in tune—it will be out of phase—which is not a good orbit. The stable orbits of an atom are the in-tune solutions and in quantum mechanics these solutions are called *wavefunctions*. The wave-fuction describes a region around the nucleus where the electron will be found. It is not a thin line, a planetary orbit. It is more like a cloud or a halo.

Back to the guitar once more. The E_2 string vibrates at 82.41 Hz. It also vibrates at 164.82 Hz, 247.230 Hz, 329.64 Hz and other frequencies. These are waves that have gone through two, three, or more oscillations when they meet on the string and merge again. So these too are stable solutions to the guitar string problem. Not only can they happen, they happen all the time in addition to the 82.41 Hz solution. These are the *harmonics* or *overtones* of this string. In fact the strength of various harmonics is how you can tell the difference between instruments. A guitar, piano and horn can all play the same note, but your ear can tell them apart because the harmonics have different strengths.

The string of a guitar also have *nodes*, places that do not vibrate. The first harmonic has half the wavelength of the fundamental frequency, so half way down the string it has a node. If you pluck the E_2 string and then put your finger at the mid-point and fret it, you will kill the fundamental (82.41 Hz) frequency, but the first harmonic will continue.

Atoms also have harmonics or overtones. There are multiple wave-function solutions to an atom. These are other orbits that are available to the electron. Much as the guitar string will vibrate with distinct harmonics, and not just any frequency, the atom also has distinct, quantized solutions or wavefunctions.

This is where things get fun. The most common wavefunction is called the *ground state*. It is a spherical region surrounding the nucleus. This is sometimes described as a shell, but that conjures up an image of a well-defined, rigid object. The wavefunctions do not have hard edges, they just fade at large distances, like a morning fog. Where a wavefunction is large is where an electron is likely to be, where it is small the electron rarely goes. So a picture of the wavefunction is like a long time-exposure photograph of the atom.

An orbit with greater energy produces a wavefunction or shell with larger radius, with smaller shells nested inside of larger ones. Like the guitar string, the higher harmonics also have nodes in them, spherical regions where the electron will not be found (see Figure 10.6).

But it gets even stranger. Because the atom is a fully three-dimensional object, its modes are more complex than the vibrational mode of a string. In fact there are wavefunctions that look like lobes and rings. The lobes and rings are tantalizing, but hard to see in nature because if you add the lobes and rings together you end up with a sphere.

Most of the time the hydrogen atom is in its ground state (Figure 10.6A). But the world is not made of just hydrogen. If you have an atom with a lot of electrons, only two can go in each wavefunction and so to

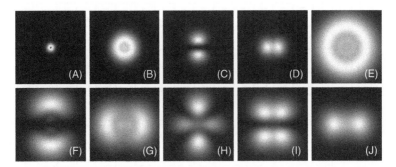

Figure 10.6 Wavefunctions of the hydrogen atom. Each image is 10 Å (1 × 10^{-9} m) wide. (A) ground state, (B–D) first excited state, (E–J) second excited state; spherical, lobes, ring, ring and lobes, double ring, ring. The rings in D, G, I and J are really two wavefunctions combined.

build an atom is to add wavefunctions on top of wavefunctions. Now we will find that the rings of one wavefunction nicely mesh with the lobes of another wavefunction, such that when you build up the atom to elements such as neon, argon and krypton, they are nicely spherical. These atoms without bumps or corners are the non-interacting, inert or noble gasses.

In non-noble gasses—the rightmost column of the periodic table in Figure 10.7—those bumps and lobes become important. That is what defines the angles of chemical bonds and the geometry of molecules.

As you add electrons we expect the atom to get bigger. But hydrogen, with one electron, has a radius of 0.53 Å, whereas helium, with two electrons, has a radius of 0.31 Å. As we have seen, you can put two electrons into a single wavefunction, which might lead you to think hydrogen and helium should have the same size. But helium also has a second proton, which pulls the electrons into a tighter orbit. Lithium is bigger than hydrogen, but as you march across the periodic table—beryllium, boron, carbon, nitrogen, oxygen, fluorine and finally neon—the atoms get smaller. Jump to the next row and we find that sodium is larger

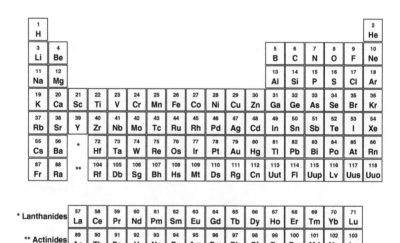

Figure 10.7 The periodic table of the elements. As electrons are added to an atom, you move across a row. Elements with completely filled shells, a smooth wavefunction, are the non-interacting noble gases, in the right-hand column.

than lithium. The trend continues through the whole periodic table. Across a row the atoms get smaller, down a column they get bigger (see Figure 10.7).

Helium is the smallest atom with a radius of 0.31 Å and cesium (used in atomic clocks), with 55 electrons, is the biggest atom with a radius of nearly 2.98 Å; that is, about six times the radius of hydrogen.

***

One last question. The heaviest element in nature is uranium and its most common isotope is ^{238}U. It is build out of 92 electrons, 92 protons and 146 neutrons. Why do we not find more massive atoms in nature? (We have synthesized a few elements that are heavier, but they tend to be unstable and short lived.) The reason atoms are not heavier is not a limit set by the electrons. It is a limit on the number of protons you can shove into a nucleus. But that is a problem of nature that takes place at a scale 100,000 times smaller than an atom, the scale of the next chapter.

***

All the things that we have looked at in this chapter—crystals, molecules and atoms—have a special type of beauty. They all have a complexity, a richness and a wide range of variations that arise from a very few axioms. Firstly, the important force holding all these things together is the electromagnetic force: opposite charges attract, like charges repel, and the force becomes weaker at a distance. Secondly, electrons are best described by wavefunctions, with all the properties of waves such as interference, resonances, harmonics and nodes. Given these two principles and a pile of electrons and protons (and neutrons) we can build all the elements of the periodic table. Given these atoms we can build all of chemistry and stitch together things like DNA or even macroscopic crystals.

The starting point is simple; that is why I say that the resulting complexity is beautiful. It is not messy and obscure. The chemistry of life, the radiant beauty of a diamond, all arise from a few simple principles, none of which surprise us.

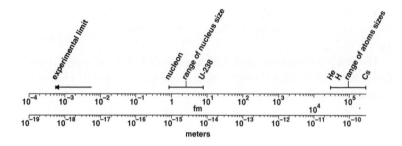

❧ 11 ❧

How Small Is Small

We ended the last chapter by looking at the way the electron's wave-function traces out the size and shape of the atom. Common atoms have a diameter of about an angstrom (10^{-10} m) and are never larger than three times that. We now continue to look at the world at smaller and smaller scales. The break between these two chapters also reflects an important boundary in nature.

What objects in nature are a tenth of the size of the atom (10^{-11} m)? Nothing. This is curious, because so far we have never discussed a scale size for which there are no naturally occurring objects or systems. So let us look a bit smaller. What is there that is one hundred times smaller than an atom? Nature entertains us with . . . nothing again. This void continues. There are no objects in nature with a size of 10^{-13} m either. Finally, the largest nucleus, the core of the heaviest natural element (uranium-238) has a diameter of about 1.5×10^{-14} m, about 10,000 times smaller than an atom.

At the very beginning of this book there was plot of a number of objects in nature laid out on a logarithmic scale. Between 10^{-10} and 10^{-14} there is nothing. Other parts of the plot are crowded and things had to be left out. But in this range nothing was omitted; rather, there really was nothing to plot. The gap must be telling us something; it jumps out at us like a missing tooth in a smile. It is significant because it marks the boundary between two great organizing principles of nature. For the moment we only comment upon its presence but will leave an explanation to the end of the book when we have all the threads in hand.

***

The next object of study is the nucleus of an atom. The nucleus is made up primarily of protons and neutrons. As we have said before, since an atom is usually electrically neutral the number of protons and electrons will be the same. So, even if it is the electrons that interact, it is the number of protons that defines an element and therefore its chemistry.

But what about the neutrons? They appear at first to have a very passive role; they do not entrap electrons. Their role also seems a bit ambiguous in so far as an element such as carbon can have a variable number of neutrons. As was mentioned when we talked about carbon-14 dating, carbon can have six, seven, or even eight neutrons. Neutrons seem to not be very rigid about their number of partners. And then there is the case of hydrogen. Most, but not all, hydrogen atoms have no neutrons. How important can the neutron be if the most common element in the universe does not need one?

The fact that hydrogen, the element with a single proton is also the only type of atom that can exist without a neutron is a clue to the neutron's role. Helium, the next simplest atom after hydrogen, has two electrons, two protons, and one to eight neutrons, but never zero. Helium usually has two neutrons, one out of a million will have one neutron, and the versions with three or more neutrons are much rarer, being artificially created and having very short lives. Elements with different number of neutrons are called isotopes. So hydrogen has three isotopes and helium has eight isotopes.

The point is that if an atom has two or more protons in its nucleus, there needs to be a neutron. If a helium nucleus had two unaccompanied protons it would be very unstable, and unstable things fall apart very quickly.

In the world of the nucleus the neutron and proton are in many ways very similar. They are about the same size and have the same mass. Neutrons and protons attract each other by the *nuclear force*. Also th clear force between two neutrons is the same as between tw even between a neutron and a proton. Therefore neutrons are often collectively called *nucleons*. The nuclear force collects nucleons (neutrons and protons) and forms them into the nucleus. But neutrons and protons are not exactly the same. Protons still have that positive charge. Two protons will attract each other via the nuclear force and also repel each other via their electromagnetic force. And so, as pointed out above, a nucleus of two protons will quickly fall apart.

Neutrons provide the extra nuclear force needed to hold things together. We can see this especially in heavier elements. With more and more protons, nature also adds more and more neutrons. For the light elements, such as oxygen or carbon, the ratio of protons to neutrons is about 1 to 1. For gold the ratio is 79:118 (about 1:1.5) and for the ultraheavy elements such as uranium it is 92:146 (about 1:1.6). This increasing ratio in the heavier elements is because the neutrons have an increasingly hard time keeping the unruly protons from pushing on each other, potentially breaking the nucleus apart.

In nuclear physics there is a chart called the *table of nuclides* (see Figures 11.1 and 11.2). It is the equivalent of the periodic table of the elements for chemistry. On it are plotted all the different isotopes. Vertically, the number of protons (or electrons) increases, so each row of the table is a new element. Starting from the bottom the rows are

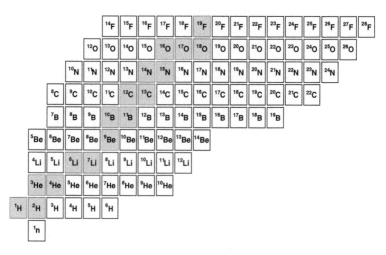

Figure 11.1 Detail from a small part of the table of nuclides. The number of neutrons increases going to the right. The number of protons increases going up. The dark cells are the stable isotopes.

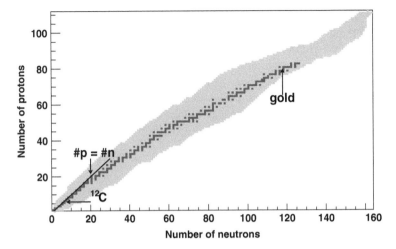

Figure 11.2 The table of nuclides. All isotopes are plotted. The dark ones are the stable isotopes. In the bottom left part of the table, the light elements tend to have about the same number of neutrons and protons. The heavier elements have 50–60% more neutrons than protons.

hydrogen (1 proton), then helium, lithium and so forth. Going across the table the number of neutrons increases. Usually each square on the table is marked with the name of the isotope. Sometimes lifetime or decay modes are included in the fine print. In isotope notation a name like ^{14}C (read as "carbon-14") tells us that there are 14 nucleons, or protons plus neutrons, and since we know that carbon has 6 protons and electrons (since it is carbon), then it must have 8 neutrons. Alternatively, take a more extreme case such as ^{238}U, uranium-238. From the table of elements, uranium has 92 electrons and protons, and so it has $238 - 92 = 146$ neutrons.

However, I really only want to talk about the big trend in this table. In the light elements there is a one-to-one neutron-to-proton ratio that nicely offsets the proton–proton repulsion. But that does not work for bigger, heavier nuclei. This is because the electromagnetic force is long range whereas the nuclear force has a short reach. In fact we can estimate the range of the nuclear force just by looking at the chart. The one-to-one ratio goes up to less than twenty protons and twenty neutrons. After that the trend becomes more horizontal; there are more neutrons then protons. A nucleus with twenty protons has a

radius of about 3–4 × 10^{-15} m, which tells us that the nuclear binding force has cut off. In reality the nuclear force is a bit shorter then
this, extending out only about 2 × 10^{-15} m, effectively only binding the
neighbor or near neighbors. This is reminiscent of chemistry. I have vi-
sions of a nucleus as a crowd, with protons as feisty characters pushing
and shoving and neutrons mixed in, trying to calm things down and
hold everything together but only talking in whispers to their nearest
neighbors.

***

I would like to "see" what a neutron or nucleus or quark looks like, but
the whole concept of seeing is going to be very different as we move to
smaller and smaller scales. But it is worth pausing a moment to dissect
what we normally mean by seeing, so we can compare it to the methods
needed to bring the subnuclear world into focus.

When you look at a maple tree in autumn and see the glowing
red leaves, what is really happening is that light of all color from the
sun has shined on the leaves. Only the light with wavelength of about
6500 Å(6.5 × 10^{-7} m) has been reflected back. These photons go in all
directions, but a tiny fraction enter your eyes and stimulate the retina.
In turn, your brain is told that a photon of about 6500 Å has arrived and
then the brain figures out that it came from an *Acer rubrum*, a red maple.
In this whole process we have a light source (sun), and object of study
(tree), detector (eye) and interpreter (brain). Laboratories where people
study particle and nuclear physics mimic these same steps.

We could not see protons or even a nucleus with classical microscopes, even if we had amazing lenses. The problem is that to see a
nucleus we need the right sort of light. Common light is the wrong
size. The light our eyes are sensitive to has wavelengths between
3800 Å(violet) and 7400 Å(red) in length. The things we are trying to
see are 10^{-15} m across, 100 million times smaller than the light waves.

If we tried to use regular light to see a nucleon it would be like trying
to see the effect of an ocean wave with a 10-m wavelength scattering off
of a virus. The wave would roll past the virus and would not be scattered
or deflected. It we had a stick a centimeter thick, a thousandth of the
size of the wave, the effect is barely perceptible. A post half a meter thick
starts to scatter waves, whereas a boulder a few meters across finally
breaks up the waves and creates interesting scattering patterns. To see

something with scattering techniques you need to match the size of the wave with the size of the object of study. To see a nucleus we need light that is 10^{-15} m across or shorter. This is what accelerators can provide.

Back in Chapter 10, when talking about the atom, we said that in a quantum system there is a wave associated with a particle such as an electron, and that if the wavelength was right the wave could wrap itself around an atom in a snug-fitting orbit. We also said that energy was related to orbit and wavelength. This is generally true for all waves, whether they are sound waves, light waves, or waves associated with an electron. In light, the longest waves are radio waves, which are not very energetic. Among the shorter waves are infrared, visible and ultraviolet, which have more energy. The shortest are X-rays and gamma rays, which have the highest energy per photon and the greatest penetration. The shorter the wavelength, the more energetic the photon. The same is true for waves associated with electrons, or any other particle. So if we want to create a wave about the size of a nucleon we can do it with an electron with about 3×10^{-11} J of energy. That does not sound like much, but it is a lot of energy for something as small as an electron. The machines that can produce these energies are called accelerators because of their ability to push particles up to extreme speeds, and so to short wavelengths.

The original accelerators worked by having one end electrically grounded and the other end at a very high voltage. A electron at one end would be repelled by the voltage at that end and attracted towards the other end. A television video tube (things that are vanishing and being replaced by flat screens) works this way, with about 20,000 V between the electron source and the screen. So when the electron hits the front of the screen it has 20,000 electron volts, abbreviated eV. The electron volt is the normal unit of energy used in atomic, nuclear and particle physics. An electron with a wavelength about the size of a nucleon has an energy of 200 million eV. This is usually written as 200 MeV or 0.2 GeV (G = giga = billion). This electron is now traveling at 99.9997% of the speed of light.

Accelerators are complex machines. The ones that use high voltages are called *Van de Graff* accelerators and can obtain energies of about 30 MeV, which is about 70 times too low to see a nucleon. They are, however, powerful enough to excite a nucleus into new configuration. However, higher voltages are very hard to control and even Van de Graff energies are approaching the level of lightning bolts. Higher-energy Van de Graff accelerators would spark, arc and randomly discharge.

Above these energies, accelerators generally push the electrons (or protons, ions, or even other charged particles) with pulses of microwaves. I used to work at an accelerator at the Massachusetts Institute of Technology. It was about 190 m long with an energy of nearly 1 GeV (one billion electron volts), which means the electrons this accelerator produced had wavelengths of about 0.2×10^{-15} m, or a quarter of the radius of a nucleon. In other words, this is about as small as an accelerator can be and just start to probe inside a nucleon, which is what we did.

***

The radius of a neutron or proton is about 0.8×10^{-15} m. In fact the distance 10^{-15} m shows up so often in these types of studies that it has been given a special name, much like the angstrom for atomic distances. It is called the *fermi*, named after the Italian–American physicist Enrico Fermi (1901–1954). If instead we had applied normal metric prefixes this distance would be called the *femtometer*, and thus by either name the abbreviation is *fm*. It was Robert Hofstadter (1915–1990), a Stanford physicist, who saw the common abbreviation and coined the term fermi in 1956. He also made one of the first measurements of the size of the proton by scattering electrons off of it. He reported its radius as 0.8 fm.

Scattering can tell us more than just the size of a nucleon. Sometimes the accelerator's beam actually knocks out a fragment of the target. We might knock a neutron or proton out of a nucleus, and then from its energy and angle deduce how deep it was packed and how tightly it was bound into that nucleus. Sometimes we knock out a new type of particle. The most common is the pion. Pions are particles in a nucleus with a unique role of holding the nucleus together. In fact there are three different types of pions, all with almost the same mass, but with different electrical charges: π^+, π^0, π^-. Pions are particles in a nucleus with a unique role. Pions are what hold the nucleus together.

So is it pions or the nuclear force that hold a bunch of protons and neutron together into a nucleus? The answer is either or both. By now we have gotten use to the idea that particles have waves; the reverse is also true. A force, usually described as a *field* can also be described as particles. Actually we have already encountered this idea before.

The particle associated with the electromagnetic force and light is the photon. To understand pions, and later in this chapter *gluons*, a quick note on the relationship between forces, particles and fields may be useful.

The simplest force to understand is the electrostatic force. Since magnetism happens when charges move, this is really the electromagnetic force, just without motion. It is the model upon which we build our vocabulary for forces. The electrostatic forces act between objects that have electric charge. The force radiates outward in all directions going on forever. It gets weaker as it radiates out as the inverse of the square of the distance: $F = \frac{1}{r^2}$. Alternatively we could say photons radiate outward in all directions going on forever. The photons become less dense as they radiate out as the inverse of the square of the distance. In other words, describing the force as a field or as particles are two sides of the same coin.

So the general features of a force are that it acts between things that have some sort of "charge." Also there are intermediate particles that travel between the principal particles. These intermediate particles are the carriers of the force and technically called "bosons." They are fleeting, without permanence. They are created by one particle, and travel to another, where they are absorbed.

Gravity is similar. Its charge is called *mass*; the greater the masses involved the greater the forces. The carrier of the force is a particle called the *graviton*. It is true that we have not isolated a lone graviton, but that is not surprising since the effects of a single one are truly minuscule. They are also expected to travel forever, like a photon.

Now we can talk about the nucleus and the nuclear force. The nuclear force may not be a fundamental force, but it does create a unique structure in nature—the nucleus—and so is most certainly worthy of our attention. All nucleons have the same nuclear charge, which was part of the reason the term nucleon was introduced. The force is carried by a pion. Now the pion is very different than either the photon or the graviton in one very important way: it has mass. To make a pion takes a lot of energy, which is borrowed from the neighboring nucleus and must be returned soon. Actually the relationship between this borrowed mass–energy and lifetime is dictated by the Heisenberg uncertainty relationship. In fact in 1935 Hideki Yukawa (1907–1981) used the range of the nuclear force and the uncertainty relationship to predict the mass of the pion. It has a mass that is about 15% of the mass of

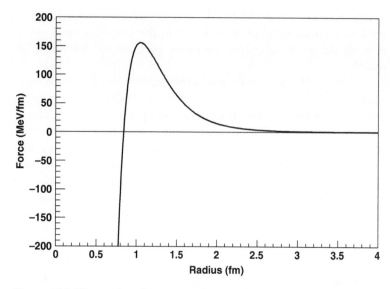

Figure 11.3 The nuclear force is repulsive at short distances, which keeps nucleons from being sucked into each other. It is attractive (positive) between 1 and 2 fm, which means nucleons are attracted to nearest neighbors. It vanishes beyond 2 fm.

the nucleons and the nuclear force has a range of only a few fermi (see Figure 11.3).

People often draw the nucleus to look like a bunch of grapes, a collection of spherical neutrons and protons clumped together, and this is about right. The nucleons like to stay about a 1 fm apart. What is missing from that image of grapes is the ethereal pions zipping back and forth. The exchange of pions bind nucleons into a nucleus much like a common currency binds an economy together.

***

What is smaller than a nucleon? The history of science has taught us to expect worlds within worlds: cells, atoms, nuclei and now what?

Inside of a nucleon are quarks and gluons (see Figure 11.4). In some sense these particles are different than anything we have encountered so far, or rather, how we know about them is much different. We have seen the tracks of protons and pions as they streak across

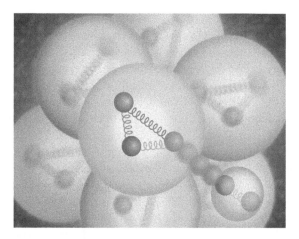

Figure 11.4 An image of a nucleus made of protons and neutrons with quarks inside of them. A nucleus appears as a collection of protons and neutrons. In the bottom right is a newly created pion. Inside of the nucleons are three quarks held together by gluons (drawn as springs). The pion contains a quark and antiquark. Courtesy of *American Scientist*, Nov/Dec 2010.

detectors leaving trails like meteors. But we have never seen a quark. We know a great deal about them, but our knowledge of these particles, which are less than 10^{-16} m across, is arrived at by indirect and subtle means.

The method works something like this. Physicists who study quarks generally are divided into two types: theorists and experimentalists. Theorists will develop models or theories about what is going on at small scales and high energies. For example they might predict that quarks gather at the center of a nucleon, or form an outer shell, or have excited orbits with a short lifetime. What makes their theories interesting is if they predict something macroscopic; if a theory makes no testable predictions it fails an important point of the scientific method.

The experimentalist will try to measure that which a theory predicts. But experimentalists are limited as to what they can measure. They can collide particles and see what comes out, at least if whatever emerges comes out far enough to reach their detectors; a particle that exists only in the atom or a nucleus will not be seen at all. That means experimentalists can look at scattered fragments that travel at least a few millimeters. They can then measure the fragment's angles, energies, mass, spins and charges, but that is about it.

Ideally, competing theories will predict different distributions of angles or energies and the experiment will pick the winner. In reality it is hard to build the ideal experiment and often the results create more questions. Still, out of decades of data we have been able to tease out an understanding of what is inside of the nucleon. The successful theories of today are all based on the idea that inside of a nucleon there are quarks, and fleeting between them are gluons. The most successful theory we have is called *quantum chromodynamics*, or QCD for short.

The force that holds quarks together, and therefore creates protons, pions, neutrons and many other particles, is the strong force, best described by QCD. What makes the strong force unique is that it has three different charges. Gravity has one charge: mass. Electromagnetism also has only one charge, even though it comes with both a plus and minus magnitude. But the strong force has three different charges. The names given to the charges are *red, green* and *blue*. They could have been named something unimaginative like q_1, q_2, q_3 or something whimsical like vanilla, chocolate and strawberry. However, red, green, blue is actually an inspired choice because it also tells us about neutrality. A particle with three quarks will have all three colors and therefore a total color-charge of *white*, which is to say neutral. Protons and neutrons are color neutral and so should have no strong force between them.

The other piece of QCD is the gluons, the intermediate particles that are the carriers of the force. Gluons can also have color-charge, which means they act very differently than photons that are neutral. Since gluons can have color they can also interact with each other. Gluons stick to gluons. Their name is not accidental. Not only do they hold quarks together, they bind gluons together into a sticky web of interactions, the result of which is that we have never seen a free quark, and probably never will.

But what if we tried to pry a quark free? What if we hit a nucleon very hard with an electron or something else? A quark can be pulled farther and farther out and the gluons will try harder and harder to contain that wayward quark. The tension in that web of gluons will rise and rise until either the quark is pulled back in or the gluon bundle snaps. When it breaks, like a rubber band snapping, more than enough energy is released to create two new particles, a quark–antiquark pair. The original nucleon gets the new quark and the escaping quark is paired with the new antiquark. The escaping pair most often is seen in the laboratory as a pion.

Once there is enough energy to create new quarks the world becomes messy or fascinating, depending upon your point of view. This is because the newly created pair might not be of the same type you started with. If you started with a proton, you could end up with a neutron, a proton, or a more exotic and rarer particle.

Quarks come in six different types, or *flavors*; and we still do not get to call them vanilla, chocolate and strawberry. Instead they are called: up, down, charmed, strange, top and bottom. The names sound whimsical, like the charges named after colors, but these flavor names are suppose to remind us that they come in pairs. The lightest and most common are the up and down quarks. The up quark is proton-like and the down quark is neutron like. A proton is made of two up quarks and one down quark, and is usually written as *uud*. The neutron is one up and two down quarks, or *udd*. The charm and strange quarks are heavier than the up and down quarks. Heavier still is the bottom quark. Finally, much heavier than any other quark is the top quark, with a mass almost 200 times greater than that of a nucleon.

QCD and the color rules tell us that a particle can have three quarks or one quark and one antiquark. Now, with six flavors one can construct 216 different combinations of three quarks, a few of which are shown in Figure 11.5. Most of these combinations are massive, exotic and short-lived. For example, the combination of up, down and strange (*uds*) may form the particle called the Λ^0, which decays into a proton and a pion in about 2×10^{-10} s. Heavier combinations disintegrate even faster. Particles that contain top quarks hardly live long enough to justify being called particles before they decay into lighter quarks. When we list these quark combinations, these exotic particles, it sounds like the whole greek alphabet: K^+, $\bar{K}^0$, Σ^-, Λ_c^+, Ω^-, Ξ_b^0, ...

This sort of physics was the bread and butter of large accelerators from the 1950s to the 1990s, when the heaviest quark, the top-quark, was finally discovered and measured. But when talking about the sizes and shapes of things in nature, what is new in the last few years is the role of the gluons. However, we have one other loose end to deal with first.

When describing extraction of a quark and creation of a pion, something that sounds like it was out of science fiction slipped into the description: antiquarks. Antiquarks are like quarks, except they are

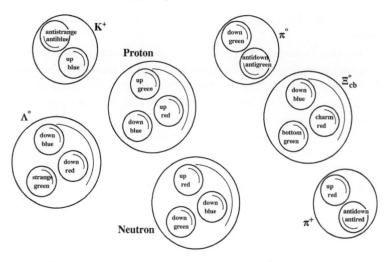

Figure 11.5 A collection of particles made of quarks. Flavors are labeled. The color charges always add up to white, or neutral.

made of antimatter. Antimatter is the stuff that is suppose to power starships or is at the heart of futuristic science fiction weapons. Actually antimatter is real; it is just very fleeting. It tends to do what the science fiction stories tell us; when it meets matter they both annihilate and release energy. It is just that in the real world these moments of annihilation take place a quark at a time.

Quarks have color: red, green and blue. Antiquarks have anticolor: antired, antigreen and antiblue. These are not new color-charges, it is just a way of writing negative color. So if we combine a quark and an antiquark with the right colors—let us say blue and antiblue—they would attract each other, and because they have opposite color charge, they would add up to white. That is in fact what is happening in a pion and we can now revisit nuclear forces, but from a quark point of view.

For example, consider a proton interacting with a neutron. In the original proton there are three quarks (*uud*). At some time, due to internal quantum fluctuations, an up quark is ejected. As it moves outward a down-antidown pair of quarks is produced, with the down quark staying in the nucleon and the antidown joining the ejected up quark. The original nucleon is now a neutron, since it contains *udd*. The pair are a pion, in this case a π^+ because of the combination of up-antidown ($u\bar{d}$) and the electric charge. Finally the pion is absorbed by a neutron, which then becomes a proton. In this whole process electric charges, color charge and flavor are preserved, as are energy and momentum. In fact the whole process moves momentum from one particle to the other, which is what keeps them bound together.

A tool for visualizing this is a Feynman diagram, named after the theorist Richard Feynman (1918–1988) who first used them. In truth, a Feynman diagram will tell us exactly the equations to solve. Here I am only using them as a cartoon. In these figures a line represents a particle. As it moves to the right, time is passing. If they are separated up and down, they are separated spatially. Line intersections or vertices are where and when particles are interacting, either creating or absorbing those intermediate particles. All of these dynamics happen within a nucleus, in fact between near neighbors, but outside of the nucleon. So Figure 11.6 shows what the world between 1 and 2 fm looks like.

We can now turn to the world inside a nucleon. Quarks are held together by gluons to form nucleons, pions and other particles. After having described the way pions hold nucleons together, the way gluons work is analogous. When we look at the world at 10^{-16} m—that is, inside a nucleon—we see three quarks swirling about each other, bound together by a mesh of gluons. Now if we closely watch the quarks and gluons we can follow their color charge, as seen in Figure 11.7. Imagine we have an up quark with a red color charge interacting with a down quark with a green color charge. The up (r) quark may emit a gluon with red-antigreen ($r\bar{g}$) color charge. That up quark is now left with a green charge. Soon the gluon is absorbed by the down (g) quark that

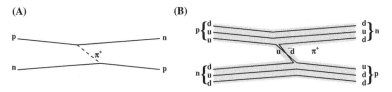

Figure 11.6 The interaction between a proton and neutron in terms of pions and quarks (A) We understand the nuclear force as the exchange of pions. (B) We can also look inside the nucleons and pions and view the force as an exchange of quarks. The gray area is just to emphasize which quarks are grouped together to form nucleons and the pion.

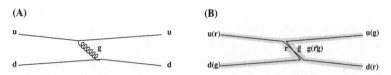

Figure 11.7 The interaction between quarks in terms of gluons and color charge. (A) Two quarks interact via a gluon. (B) A red quark could emit a red-antigreen gluon and then becomes green. That gluon could then be absorbed by a green quark, which then will have a red color charge.

now takes the red color charge. At all times color charge is conserved, and therefore the quarks are bound.

So let us step back once again for a moment. Inside of a nucleon there are always at least three quarks. The quarks are buzzing around like electrons in an atom. Between them gluons are shooting back and forth. By the fact that gluons have color charge the gluons can interact with not only quarks, but with other gluons. There is also enough energy for quark–antiquark pairs to be produced all the time, sometimes with exotic heavy quark flavors. And every once in a while a pion will pop off the surface of the nucleon and be absorbed by a neighbor. I imagine it like a boiling pot of tomato sauce that generally stays in the pot; occasionally a small droplet will spatter out and cover the stove top with red. Nucleons contain a whirl of activity.

We have only touched the tip of the iceberg. There are a great number of combinations of flavors and colors as well as spins, antiquarks and excited states. Also it has become more and more apparent that gluons are not just the cement that holds quarks together, but that they also have an important role in creating the macroscopic particles and determining the properties of those particles. But in this book we are concentrating upon the sizes and shapes of things. So I will only add one more tidbit about neutrons.

Two down quarks and one up quark swirl about inside a neutron. The question is how do they orbit each other? We would expect all three to occupy an orbit much like the ground state of the atom. It will be spherically symmetric, only much smaller, since the strong force binds them tightly. At first we would expect all three quarks to be on top of each other. However, there is a small correction. The two down quarks end up spin-aligned which gives them a slight magnetic repulsion, pushing them out. At the same time the up quark is magnetically attracted

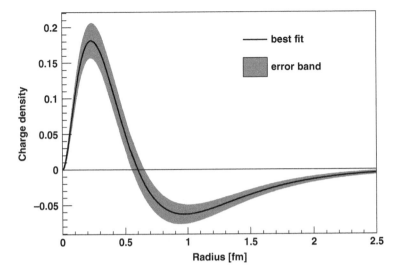

Figure 11.8 The electric charge distribution of a neutron. The total electric charge is zero – a neutron is neutral. However, the outside is negative and the inside is positive. Courtesy of the BLAST Collaboration/MIT Bates Laboratory.

towards the other two, pushing it towards the middle. These are small changes in their configuration, with the up quark's orbit radius just under 0.4 fm and the two down quarks at just over 0.4 fm. This means that the outside of a neutron is slightly negative and the inside is a bit positive (see Figure 11.8).

***

The biggest accelerator on Earth, the Large Hadron Collider at CERN (*Organisation Européenne pour la Recherche Nucléaire* or European Organization for Nuclear Research), is located under the Swiss–French border near Geneva. It has an energy of about 8 TeV (T = tera = 10^{12}). That means that the particles in the beam have a wavelength of about 5×10^{-19} m. This means that our present experimental limit is about 10^{-18} m. So when we look at the world at that scale what types of structures do we see? Actually the smallest structure we have seen is the orbits of the quarks in the neutron and that was observed at 10^{-16} m. These high-energy machines are not looking for structure as much as for new forms of matter: heavy quarks, Higgs bosons and so forth.

So our search downward ends at 10^{-18} m; we have hit bottom. But we have not measured such obvious things as the size of an electron or a quark. In fact, according to QCD, quarks and electrons are *point-like*. Their diameters are zero. That is something hard to think about. These are not dots that shrink until they are gone, because they are still there. Trying to imagine something with a zero radius is as taxing as imaging something with an infinite radius. But are they really point-like? We do not know and can only say that from an experimental point of view they are smaller than 10^{-18} m. We also do not know if QCD is the last word in particle physics. It is very successful: it explains most of particle physics and probably most of nuclear physics as well. However, we are still allowed to ask the question, "What might be smaller?" A dimension of 10^{-18} m only marks the line between what we know and what we suspect. Physics does put limits upon what happens at smaller scales, but without sharp details. And we, being humans, still ask the questions.

I made the statement that according to QCD quarks and electrons are point-like and that all experiment so far supports that. But this is like standing at the end of the Pan-American highway in Prudhoe Bay, Alaska, and saying, "Based upon what I can see, Alaska goes on forever." Alaska is huge, but it does not go on forever. Electrons and quarks are tiny and must certainly be smaller than 10^{-18} m across for QCD to be so successful, but that does not mean that there is nothing else smaller in nature.

The *standard model* of particle physics is our best description of nature at this scale. It combines our best explanation of the strong, weak and electromagnetic forces, so QCD is one of its components. The standard model incorporates everything particle physicists think about except gravity. The outstanding question in the field over the last decade is not whether this model works or is right, but rather whether it is the last word in particle physics.

I was once at a talk about the problems of the standard model—these talks are almost always called "Physics Beyond the standard model'—when a senior professor from MIT leaned over and whispered to me, "The problem with the standard model is that there are no problems with the standard model." People have spent years looking for holes in it and the best we can say is that it does not feel right. It is too gangly, too unwieldy; it has too many parts; it lacks elegance and aesthetics.

And then there is gravity. The standard model does not even attempt to explain gravity. Also, our best theory of gravity—general relativity—cannot simply be appended to the standard model. There are a number of candidate theories and models for higher energy, smaller scales and more exotic particles, but at this time we do not have the data or tools to move forward with certainty.

***

Instead of trying to see clearly the next smallest structure (if there is one), we can try to turn the problem around and ask that very fundamental question, "Is there a lower limit of nature?" Surprisingly enough, if quantum mechanics is still the law of nature at ever-shrinking scales, there *is* a bottom. There is the Planck length.

We mentioned the Planck length in Chapter 5 when describing the Planck constant, but only briefly. Max Planck proposed the Planck length as

$$l_P = \sqrt{\frac{\hbar G}{c^3}} = \sqrt{\frac{hG}{2\pi c^3}} = 1.6 \times 10^{-35} \text{ m}$$

It is a combination of the Planck constant (h or $\hbar$), which is at the heart of quantum mechanics, Newton's gravitational constant (G), which is the key to both Newton's law of gravity and Einstein's theory of general relativity, and the speed of light (c), famous from special relativity. Anyone, anywhere in the universe could measure these three constants and construct the Planck length. It is truly universal. It does not depend upon the length of a platinum–iridium bar in Paris or the size of our home planet or the choice of a spectral line. The Planck length is set by nature, not by mutual agreement, which makes it unique. But is it significant?

Consider a *Gedanken*: a thought experiment. If I am trying to interact with a particle that is a Planck length across I will need a photon with a wavelength of about a Planck length. A photon with that wavelength will have an energy of about 10^{28} eV or 0.1 J. That is not a lot of energy. But when it is crammed into such a tiny particle it would make it into a black hole, which will swallow up any photon that is trying to escape. In other words, if there is a particle as small as the Planck length it will be invisible. We can never, even in principle, interact with it.

***

One other important feature of the Planck length is that this is the scale of the world where *string theory* operates. String theory was created to deal with a number of problems in particle physics theories. One of the first problems was the fact that in most theories forces become infinite as distances go to zero. We have said that the forces of gravity and electromagnetism decrease as $1/r^2$ as the distances become large. But that also means they increase as the separation between charges approach zero. In fact, at zero the force would be infinite $\left(\frac{1}{0^2} = \infty\right)$. Mathematically this is called a *singularity*. Physically it is unpalatable. String theory says that at a very small scale particles are not truly point-like, rather they are loops, described as *strings*. It also suggests that different types of particles may just be the same loop of string oscillating in different modes. That would help the standard model by explaining all of its parts as just simple modes of oscillation. But string theory has yet to get to that stage.

It has also been proposed that not only particles, but the very nature of space and time are different on this scale. Instead of space and time as a continuous fabric, it is sometimes described a *space-time foam*.

I like the motivation behind string theory, but to me it has one major difficulty. After several decades of development it has yet to make any unique and testable predictions. Actually, that is not really surprising when you remember that the size of these strings are ten quadrillion (10^{16}) times smaller than the scales probed by our biggest experiments. This is like starting with quarks and QCD and trying to figure out which way a butterfly will flutter. It may work in principle, but it may be very difficult in practice.

∗∗∗

When Johann Loschmidt first measured the size of an atom in 1865 he wrote:

An imposing string of numbers such as our calculation yields, especially when taken into three dimensions, means that it is not too much to say that they are the true residue of the expectations created when microscopists have stood at the edge of the bottomless precipice and described them as "infinitesimally small." It even raises the concern that the whole theory, at least in its present form, might wreck on the reef that the infinitesimally small cannot be confined by experiment.

When Loschmidt looked into the bottomless precipice he could see ten orders of magnitude. We can now see nearly twice that. Yet if the true bottom is down there at 35 orders of magnitude we might never clearly see it. However, maybe there is something between 10^{-18} and 10^{-35} m. That is a large region and nature may yet hold more twists and surprises. Or maybe space and time at the Planck scale are not quite what we think they are.

We stand at the edge of the precipice with Loschmidt and wonder.

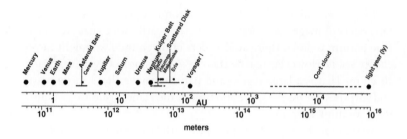

❦ 12 ❦

Stepping Into Space: the Scales of the Solar System

The night of 7 January 1610, was clear above the city of Padua, in northern Italy. Galileo Galilei was out that night looking at the stars. Later he wrote about that evening, "And since I had prepared for myself a superlative instrument, I saw (which earlier had not happened because of the weakness of the other instruments) that three little stars were positioned near [Jupiter]." With his simple hand-ground lens and homemade telescope, the bright disk of Jupiter would have been distorted and the colors would bleed, with white separated into blues and reds. But what caught his attention was that these three new stars formed a straight line, two on the east side of Jupiter and one on the west. The orientation of his diagram, with east (*Ori.*) to the left and west (*Occ.*) to the right seems backwards at first. But he was drawing a map of the sky overhead and not a map of the land underfoot.

Galileo also noted that this line of stars were parallel to the plane of the eclipse, which is the path of the planets as they move across the sky. But to him, on that night, this was just a coincidence.

". . . guided by I know not what fate . . ." Galileo pointed his spyglass again at Jupiter the next night and saw all three stars to the right.

Ori.   O   *   *   *   **Occ.** Jan 8, 1610

At this time Galileo assumed that the newly seen, tiny stars were fixed in the sky like all other stars. So he interpreted the rearrangement of these stars relative to Jupiter to mean that Jupiter was moving to the left, or east. But that was contrary to the calculated motion of the planet for that day.

On 9 January Galileo was prepared to look at Jupiter again, for what he had observed did not fit into his understanding of planetary motion. But the sky was cloudy and did not cooperate. I have always liked that night's report. For weather is very real, and real stories, real discoveries can be interrupted by very real clouds.

The next night, 10 January, was again clear over Padua and again Galileo trained his telescope upon the great planet. This time there were two of his new "stars" to the left of Jupiter. Galileo thought that the third may have been hidden behind the planet. The mystery was deepening, but Galileo was also starting to see a possible interpretation: ". . . now, moving from doubt to astonishment, I found that the observed change was not in Jupiter but in the said stars."

Ori.   *   *   O   **Occ.** Jan 10, 1610

Galileo knew that he was seeing something important and so he started making a series of meticulous observations. A few months later, in March 1610, he published his results in a book, *Sidereus Nuncius* or *Starry Messenger*. In this book he includes 65 sketches of Jupiter and its satellites, whereas in the rest of the book he has only five diagrams of our moon, two of star clusters, and two of nebulae. That is because there was something unusual, even extraordinary, going on around Jupiter. His drawings became more and more detailed, with one "star" soon marked as larger. On 12 January, while he was watching at 3:00 in the morning, one of the stars came out from behind Jupiter. On the next night he spotted a fourth "star."

Galileo soon realized that these were not stars. As Jupiter traveled across the winter sky the four companions followed. His later sketches even marked "fixed" stars in the background, as Jupiter and its entourage drifted across the sky. Within a year, Johannes Kepler had read

Galileo and coined a new word for these companions; his word *satellite* is based on the Latin word for "attendant."

***

There are a lot of good stories about the "Galilean moons." The reason I wanted to introduce them, and Galileo, is that so many of the puzzles and solutions encountered in trying to figure out the size of the solar system are illustrated by these four satellites. At about the same time as Galileo was playing with his telescope, Johannes Kepler was deriving his *laws of planetary motion*. These laws apply to satellites like Jupiter's as well as planets. They gave a relationship between orbit periods and size. Orbit periods are easy to measure, but you also need one distance in the solar system to set the scale for everything. If we knew the distance to the Sun, we would know the size of the orbits of all the planets and the size of the solar system.

Our best measurement of the Sun–Earth distance are based upon knowing the speed of light, and the first good measurement of the speed of light was based upon the Sun–Earth distance and the moons of Jupiter. In fact one of the first attempts to measure the speed of light was also made by Galileo.

***

At one time René Descartes, a contemporary of Galileo's, said, "I confess that I know nothing of Philosophy, if the light of the Sun is not transmitted to our Eyes in an instant." Does light have a finite speed, or does it travel instantly? This had been an open question for centuries. But it was Galileo, more the experimentalist than philosopher, who first tried to bring the question into the laboratory, or at least out of the philosopher's salon.

In his book *Dialogues Concerning Two New Sciences* (1638) he proposed a method. To start with, two people, each with a lantern with a shutter face each other. Initially the shutters are closed, to cut off the lanterns' light. Then the first person opens their shutter. When the second person sees the light, they unshutter the second lantern. The first person will note the delay time between when they opened their shutter and when they saw the other lantern's light. Galileo reasoned that at short distances most of the delay was due to the reaction time of the two people. The measurement was then repeated over a much longer distance with any additional time being due to the time it takes light to travel between them. Galileo reported, in the voice of Salviati:

In fact I have tried the experiment only at a short distance, less than a mile, from which I have not been able to ascertain with certainty whether the appearance of the opposite light was instantaneous or not; but if not instantaneous it is extraordinarily rapid – I should call it momentary; and for the present I should compare it to motion which we see in the lightning flash between clouds eight or ten miles distant from us.

***

The next attempt to measure the speed of light takes us back to watching the moon of Jupiter. Within a year of observing those moons, Galileo realized that there was a potential to use them as a great universal clock hung in the sky for everyone to see. Or at least it offered a way of synchronizing clocks. For instance, if the observatory at Paris or Greenwich announced that Io, one of those moons, would slip behind Jupiter at exactly ten minutes after midnight, the whole world could watch the event, the *occultation*, and synchronize their clocks. This would have beneficial consequence for navigation and commerce.

One of the great technical challenges of the seventeenth century was how to measure longitude, especially at sea. A mariner could measure the elevation of the Sun above the horizon and then by knowing the tilt of the Earth on that day, they could calculate their latitude. But longitude was much more difficult. A solution that people understood was based on knowing local and standard time. If you know the time at a standard location, such as the observatory at Greenwich, and you can determine your local time, you can then calculate your longitude. For example if one day you know it is local noon, since the Sun is due south (with corrections based on the equation of time or the analemma) and at the same moment it is 9 AM in Greenwich, you know that you are 3 h before Greenwich. You are an eighth of the way around the world to the east. You are at 45° east, which is the longitudinal line that runs near Volgograd in Russia (48°42′N, 44°31′E) or Baghdad (33°20′N, 44°25′E). The problem with this technique was that no seventeenth century clock was reliable enough to tell you the time in Greenwich, especially as ships would have to carry these clocks and would be bobbing on the ocean for months or even years. If you are at the equator, a clock that is off by a single minute means your longitude calculation is off by 27 km (17 miles). If you are off by an hour, you are in the wrong time zone.

Galileo, and many other astronomers, realized that if someone could hand you a chart or table that listed the moment that Io was eclipsed by Jupiter as seen in a standard location, you could synchronize your clock. You could be sailing in the Indian Ocean, or the South Pacific, and you would see that event exactly at the same time as it was seen in Padua or Greenwich or Paris. And then you could determine your longitude.

Galileo worked for years trying to devise such a table, but never succeeded; the orbits did not seem to be regular enough. He also was not the only one to recognize the utility of such a table. In fact the Paris Observatory and the observatory at Greenwich were started in 1667 and 1674 in part to create such a table. As these countries sought to build far-flung empires and rule the waves, it was a concern of national importance and pride and so deserved public financial support.

The first director of the Paris Observatory, which is still located at the southern tip of the *Jardin du Luxembourg* (Luxembourg Gardens), was Giovanni Domenico Cassini (1625–1712). Cassini was a great Italian astronomer and it must have been quite the coup for the organizers of the observatory to get him to come north. In fact he made so many discoveries, especially concerning Saturn, that a recent space probe sent to that planet was named after him.

Cassini had a young Danish assistant, Olaf Rømer (1644–1710) who was set to work on the moons of Jupiter problem. Most of the work concentrated on Io, whose period would appear to get shorter for half the year, and then longer. Rømer realized that it got shorter as the Earth approached Jupiter and longer as we backed away, and so in 1675 he suggested an audacious solution. He proposed that light from Io and Jupiter took longer to reach the Earth the farther away they were. We do not see the eclipse at the moment Io passes behind Jupiter. Instead we see it a few minutes later when light from the event has crossed the millions of black leagues between us. This meant that light took time to travel.

Rømer, upon examining all the data that he and Cassini had collected, calculated that it took about 22 min for light to travel across the Earth's orbit. Actually this time was later corrected to a value closer to the $16\frac{1}{2}$ min we measure today. This was combined with a measurement of the Earth–Sun distance and a speed of 2.75×10^8 m/s was determined, only about 8% slower than modern measurements. The problem was not with the reasoning or the time, but with the fact that the diameter of the Earth's orbit was not known very well.

The distance from the Sun to the Earth is called an *astronomical unit* (AU), and it becomes the basis upon which all astronomical measurements are made, because it is our baseline for surveying the stars and planets. But it is not an easy measurement and things had not changed much since the time of the Greeks until the middle of the seventeenth century.

In 1672 Cassini sent the Jesuit astronomer Jean Richer to Cayenne in South America. Cassini and Richer both measured the position of Mars at the same time. They had to do it at the same moment because the Earth is in motion, both spinning and orbiting. After a long voyage and a delicate measurement, they derived a value of 138 million km (compared to a modern measurement of 150 million km), which is what Rømer used.

***

The importance of the astronomical unit really came to light a hundred years before Galileo. When Nicolaus Copernicus (1473–1543) was putting together his model of the solar system he could not see details of the planets: he preceded the telescope by a century. But he could measure angles well. He could plot the paths of the planets in the sky and fit them to his ideas about circular orbits. He even published estimates of the radius of the planets' orbits in terms of the AU, which agreed remarkably well with modern numbers (see Table 12.1). The

Table 12.1 Radius of the orbits of the planets: Copernicus vs modern measurements.

Planet	Symbol	Copernicus AU	Modern AU
Mercury	☿	0.38	0.387
Venus	♀	0.72	0.723
Earth	⊕	1.0	1.000
Mars	♂	1.52	1.524
Jupiter	♃	5.22	5.203
Saturn	♄	9.17	9.537

fact that Copernicus got the Earth's orbit precisely right is not surprising. An AU was and is defined as the distance between the Sun and the Earth. But Copernicus could go no farther. He knew the size of the other planet's orbits in terms of the AU, but he did not know how far an AU was in terms of distances used on Earth, like the mile or the meter.

The AU shows up in Kepler's third law of planetary motion. In that law he states that the orbital time of a planet (T = period) squared, is equal to the radius (more precisely, the semi-major axis) of the orbit cubed

$$T^2 = r^3 .$$

That means that if you could measure a planets orbit time in years, you could calculate its orbital radius in AU. Kepler not only could check this against the planets but since this law emerged after 1610, he could also check it against the satellites of Jupiter (see Figure 12.1). If only he could stretch a steel tape measure from here to the Sun, he could map the whole solar system.

So Copernicus and Kepler reported the size of the orbits of the planets in terms of the AU, but the AU—the baseline of astronomical measurement—was elusive. Cassini and Richer measured by observing Mars from Paris and Cayenne, using the width of the Atlantic as a baseline, but their estimate had a lot of uncertainty. Cassini and Rømer made another measurement. This time they both stayed in Paris, but they let the Earth's rotation move them to a second location. It was a simple measurement to perform, but a difficult one to analyze. But the most important measurement of the AU was the *transit of Venus*, made a century later in 1769. It was an event with which the names Captain Cook and Point Venus in Tahiti are forever linked.

***

The transit of Venus measurement was conceived of by Edmond Halley. Like Halley's comet, the transit of Venus is a very rarer event. A transit is when Venus (or Mars) passes in front of the Sun, appearing from Earth as a black spot in front of the glowing orb. It is rare because the orbits of the planets are not in the same plane. In fact it only happens if the Earth and Venus both cross the line where their planes intersect, an alignment that happens only about once every 130 years, and then again eight years later. After Halley conceived of the measurement the

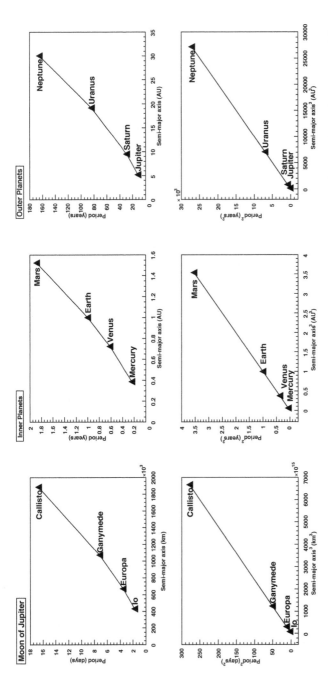

Figure 12.1 Kepler's third law demonstrated by Jupiter's moons and the planets. The top row shows orbit radius vs period for Jupiter's moons, the inner planets and the outer planets. The second row shows the square of the period versus the cube of the radius for these three cases. Since the moons and planets lay on a straight line this means $T^2 = r^3$.

next transits would be in 1761 and 1769, then again in 1874 and 1882, and recently in 2004 and 2012.

The reason there was such optimism for this technique is that it depended upon the measurement of time instead of angles. The amount of time it takes for Venus to transit the Sun depends upon the distance to the Sun and your position on Earth. For instance, two observers 10,000 km apart on Earth will see Venus crossing slightly different parts of the Sun. The transit time they measure will typically differ by a half hour or so (see Figure 12.2). A remote observer with a moderately accurate watch can measure the transit to within a second, which should lead to a distance measurement error of less than 1%. Halley had measured the transit of Mars in 1677 and felt that with a coordinated effort the AU could be determined to 0.2%.

Halley did not live to see either of his predictions confirmed, but his comet returned in 1758–9 and in 1761 and 1769 the scientific community of Europe mounted a worldwide effort to observe the Transit of Venus. The story of that measurement is a heroic adventure, because you need people to trek across the Earth to widely spaced positions to obtain the greatest possible baseline. Expeditions were mounted by the French Academy of Science, the British Royal Society, as well as numerous individuals across the globe. Captain James Cook with the astronomer Charles Green went to Tahiti, where the sailors nearly traded away the nails that held together their ship, the HM Bark *Endeavour*.

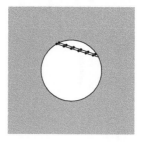

Figure 12.2 The transit of Venus as seen by two separate observers. Two observers, 10,000 km apart on the Earth, will see Venus crossing a different part of the Sun. The black spots show Venus every hour during the transit. One observer sees the transit in just over 5 h, the other in just under 6 h. In this case the difference was a half hour, and with an uncertainty of 1 s that is one part in 1800.

Father Maximilian Hell travelled above the Arctic circle to Vardø in Norway. William Wales and Joseph Dymond spent the preceding winter at Churchill on the Hudson Bay in Canada, waiting for the great event, watching the thermometer drop to $-43°$F and seeing brandy freeze.

One of the most disappointing expeditions was that of Guillaume Le Gentil, who traveled to Pondicherry in India. His travels were interrupted by war between England and France and so he was afloat for the 1761 transit. He decided to stay there until 1769 and was ashore and prepared for the second encounter, but that morning the Sun was obscured by clouds. When he returned to France he had been gone so long that he had been declared legally dead, his wife had remarried, he had lost his seat at the French Academy, and his estate had been plundered by relatives.

Abbé Chappe d'Auteroche travelled across Russia on the "highway" of the frozen Volga River to Tobolsk in Siberia for the 1761 transit. Then, eight years later he traveled to San José del Cabo, on the Baja California peninsula of Mexico. Here he made a detailed measurement, but unfortunately everyone except one person in his expedition died of "epidemical distemper." However, his notes and measurements still made it back to Paris.

These were the most extraordinary efforts and took place in the midst of England and France struggling for world domination. The Seven Year War (1756–1763), or the French and Indian War, as it is known in North America, was in full fury. Yet there was a unique camaraderie among scientists internationally. The Royal Society secured letters of passage for French astronomers in the war zones and the French Academy of Science reciprocated. Data from both empires and dozens of other places was collected and shared. In the end the final and best analysis was done in Germany. Johann Encke of the Seeberg Observatory in Gotha published his analysis in 1824. Among his results were a number of new longitudes, because you need to know where you are when making these measurements. He reported a solar parallax of 8.58 arc-seconds and a distance of $153, 340, 000 \pm 660, 000$ km. That reported uncertainty was just under half a percent. In reality his estimates were a bit more than 2% off.

The solar parallax is the angle that the Sun would appear to shift in the sky if you instantly moved from the Earth's equator to a pole. An arc second is 1/60th of an arc-minute, and an arc-minute is 1/60th of a degree. So 8.58 arc-seconds is exceedingly small. It is the angle between two pins in a cork that are 4 mm apart and viewed from 100 m away.

Scientists did not achieve the 0.2% that Halley had hoped for primarily due to a phenomena called the *black drop effect*. As Venus approached the edge of the Sun, the edge seemed to bleed into the Sun, making the moment of contact and the start of the transit hard to identify. This meant that the exact second of contact could generally not be determined.

Of course, the expeditions of 1761 and 1769 were not the last word on the astronomical unit. In 1874 and 1882 another worldwide effort was mounted. This time the United States was a major participant. There was also a widespread use of photography. This meant that photographs could be taken as Venus approached the Sun's edge, and even if the moment of contact was still blurred by the block drop effect, astronomers could deduce that moment by extrapolating between photographic images. With the aid of trains and steam ships the expeditions took less time and the data returned faster than in the previous century. Their final result was a figure of 149,840,000 km. This is about a 0.2% difference from modern measurements, a result that I think would have satisfied Halley.

***

There was a recent pair of transits in 2004 and 2012, but they were more curiosities than important measurements. Our best measurements of the AU now come from using radar. We bounce radio waves off of nearby planets, Mars, Venus and Mercury, and time how long it takes until the echo is received. Ironically, instead of measuring the distance to the Sun and using Kepler's laws to get the distance to the planets, we do it the other way around. Radar off of a rocky planet is a cleaner signal than off of a gaseous planet, so we use Mercury, Venus and Mars. Finally, our best number for the AU is $149,597,870.691 \pm 0.030$ km. That means that we actually know the AU to ten digits of accuracy, and to within 30 m.

In some sense the way we measure the distance to the planets and the Sun now seems too easy. We do not mount expeditions to far-off continents, risking life and limb and taking years. All we do now is bounce radio waves off of Venus or Mercury and time how long it takes for the signals to return. But we stand on the shoulders of giants. The technique only works because we know the speed of light so very well, and that number has also taken centuries to establish.

***

Once we have measured the AU and the size of the Sun, the dimensions of planets and various satellites are much easier to determine. To measure the size of the orbit of Mars or Uranus, we need only measure its orbit time and apply Kepler's third law. If we know how far away Neptune is, and measure the angle it subtends in our telescope, we know its size. So now we can march out to the edge of the solar system, naming the sizes of things as we go. But if we did that we would miss a lot of interesting stuff. Instead I want to view the sizes of the planet's orbits as a frame, a canvas upon which we can now go back in sketch in details. I want to look at the sizes and shapes of planets, moons and asteroids.

***

One of the most striking things about the solar system is that the Sun, the planets and many of the moons are spherical. In space it is so common that it is hard to think that it could be otherwise. All the large objects, the things that catch your eyes, are spherical. But here on the surface of Earth it is a rare natural object that is spherical. Sometime fluids, like water droplets, may be spheres but not rigid bodies.

Some of the natural satellites, or moons, of the planets are spherical and some are not (see Figure 12.3). At present there are over 180 natural satellites catalogued in our solar system, and that number keeps

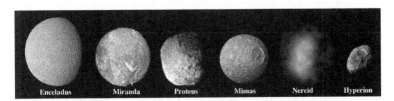

Figure 12.3 Some of the moons in our solar system. The images are arranged by size, from the spherical Enceladus (radius 252 km) to the irregular Hyperion (radius 133 km). Nereid orbits Neptune and is difficult to photograph. Courtesy of NASA.

growing. Of all the moons, 19 are spherical. For example, the two moons of Mars, Phobos and Demos (named after the sons of Mars from mythology), are irregular lumps of rock, reminding one of potatoes with dimples and pock marks. Moons such as Prometheus and Calypso of Saturn are even elongated. And then there are the asteroids, most of which are very far from spherical. So why is it that some are globes and others are potatoes? Why spherical or irregular?

The trend here is that the larger the moon, the more likely it is to be spherical. All the moons larger than, and including, Miranda are spherical. Miranda is a moon of Uranus and has a radius of 236 km. All the moons smaller than Nereid are irregular. Nereid orbits Neptune and has a radius of 170 km. Between these two are Proteus, which is irregular and has a radius of 210 km, and Mimas, which is spherical with a radius of 200 km. There really is a threshold somewhere about 200 km, above which moons form into spheres. To understand this, we will first look at the two extremes.

A star, which is clearly spherical, is held together by gravity. A boulder or cobble need not be spherical, and is held together by chemical bonds. We have now seen that chemical bonds are a result of the electromagnetic force. So at first we might reason that an object shaped by gravity or the electromagnetic force would have the same shape since both forces have the same form. They extend over all space and get weaker as $1/r^2$. Even though the electromagnetic force is so much stronger, the fact that it comes with both a positive and negative charge means that Calypso can have an odd shape.

We have already discussed chemical bonds in Chapter 10 and also the fact that electrically neutral atoms can still feel electromagnetic forces when they are close to each other because one part of the atom may be more positive and one part more negative. That is the source of chemical bonds, and that means that by their very nature chemical bonds are short ranged and essentially work only between neighbors.

A rough rock that I might pry out of a cliff face is jagged and stays jagged. This is because of the powerful chemical bonds between the atoms. Each atom is tightly bound to its neighbor. But all the atoms are ignorant of any distant atom, where "distant" can mean a few angstroms away.

Gravity, by contrast, is the great shaper of the cosmos and it is a lot different than the electromagnetic force in two major ways. Firstly, it is a lot weaker. Secondly, the charge is mass and only comes in one type. Mass attracts mass and there is no "anti-mass" that can repel or neutralize that attraction. Therefore an atom on one side of the moon gravitationally interacts with—and attracts—atoms on the other side of the moon. The bigger the object, the more mass, the greater the forces that pull that body into a sphere. At some size one crosses a threshold. The global gravitational forces overwhelm the local chemical forces and reshape that body into a sphere.

There is in fact, no simple threshold. We cannot derive a line between Miranda and Proteus, the spherical and the irregular, by simply balancing chemical and gravitational forces. If our Moon was split in half, each piece would be big enough to eventually form new spherical moons. But it would not reform into spheres in an instant, and it would never be a perfect sphere. Even on our home planet, geology is raising up mountains, whereas erosion and landslides (gravity), are pulling them down. Still, given enough time and enough mass, a large irregular moon or asteroid will become more spherical.

***

One other place where these two forces compete is something called the *Roche limit* or *Roche radius*. Named after the French astronomer Édouard Roche (1820–1883), it was inspired by the observation that the rings of Saturn are inside the orbits of the spherical moons (see Figure 12.4) Since Roche's days we have observed that his limit also applies to all plants with rings: Jupiter, Saturn, Uranus and Neptune. After his observation, Roche calculated the forces within a moon. He balanced the tidal forces, which distort, against the gravitational forces, which shape a moon into a sphere. He found that if a moon was to venture too close to its *primary* (the planet it is orbiting), the tidal forces from the primary will exceed the moon's self-cohesion and the moon would break-up. Roche proposed that this is how the rings of Saturn were formed.

Within the Roche limit you can have satellites, but they are rigid; that is, they are held together by their chemical bonds and not gravity. For example, the Roche limit for the Earth is between 9,000 and 18,000 km, depending upon the rigidity of the satellite. Our moon is well beyond that limit, orbiting at about 400,000 km; it is safe from tidal break-up. On the other hand, man-made satellites, such as

Figure 12.4 The rings of Saturn. A moon within the Roche limit would break up due to tidal forces. Courtesy of NASA.

the International Space Station and the Hubble telescope, which are between 400 and 600 km up or a bit less than 7000 km from the center of the Earth, are well within the Roche limit. But they are held together primarily by chemical bonds and not gravity.

***

Whereas much of the structure of the solar system is determined by the balance between chemical bonds and gravity, other aspects are governed by history and the initial conditions of the region of space where our solar system evolved. Thus there are parts of our description that are unique to the solar system and parts that we can expect to find in any planetary system, even around stars in distant galaxies.

The size of the bodies in our system are a result of how much gas and dust was in the region when the Sun and planets were forming, about four and a half billion years ago. This region of space was then filled with material that must have included the ashes of a supernova, since it contained elements heavier than iron. Gold in a wedding band is proof that the Earth contains remnants of older stars. The gasses and dust in the embryonic nebula attracted each other due to gravity, and they started to coalesce. Once the core of this cloud was dense enough the Sun formed and started to burn. This young star radiated a powerful solar wind that pushed a lot of the light elements out and left the heavier elements near to the Sun, in the inner solar system. So the planets that formed close to the Sun contain rocks and metals. These are the *terrestrial planets*: Mercury, Venus, Earth and Mars. Farther out, the gasses or *Jovian planets* formed: Jupiter, Saturn, Uranus and Neptune.

Planets range in size from Mercury, with a radius of 2440 km, to Jupiter, with a radius of 69,920 km (see Table 12.2). This observation leads us to the question of whether planets could be of a different size? At the lower end, a planet cannot be much smaller than Miranda if it is going to be spherical. At the other end we think a planet cannot be more than 40% bigger than Jupiter. We will see why in the next chapter. In the last few years we have started to observe *exoplanets*, planets that orbit other stars. We wait to see if our theories, based upon what we see with our eight local planets, holds true when we measure hundreds of new planets.

***

So now we can start our tour of the solar system, starting with the inner planets and working our way out to the cold, dark regions of space.

One curious trend that was noted over 200 years ago is that not only does the spacing between planets increase as you move out from the

Table 12.2 Some properties of the planets.

Planet		Distance from Sun ($\times 10^6$ m)	(AU)	Radius ($\times 10^6$ m)	Mass ($\times 10^{24}$ kg)	Period (years)
Mercury	☿	57.909	0.387	2.439	0.3302	0.24
Venus	♀	108.208	0.723	6.051	4.869	0.62
Earth	⊕	149.597	1.000	6.378	5.9742	1.00
Mars	♂	227.936	1.524	3.397	0.6419	1.88
Jupiter	♃	778.412	5.203	71.492	1,898.7	11.86
Saturn	♄	1,426.725	9.537	60.267	568.51	29.45
Uranus	♅	2,870.972	19.191	25.557	86.849	84.02
Neptune	♆	4,498.252	30.069	24.766	102.44	164.79

Sun, but it seems to do so in a regular pattern. The trend was quantified as the Titus-Bode law:

$$r = \left(\frac{4 + 3n}{10}\right) \text{AU} \quad n = 0, 1, 2, 4, 8, 16, \ldots$$

where n doubles beyond 1. The law is also illustrated in Figure 12.5. It is just an observational trend, which fits the data pretty well, except for the exceptions. Where is the fifth Titus-Bode planet? When Uranus was discovered it fell where number eight had been forecast. However, Neptune is not near any predictions (although Pluto was also close). The equation was formulated to fit the data and has no deep theoretical basis. However, it may be more than just complete coincidence.

There is something called *orbital resonance*. Two planets that are orbiting the Sun will push and pull on each other every time they pass each other. If two orbits are in resonance, the smaller planet can be pushed out. A way to see this is to imagine pushing a child on a swing. If the period of the push and the period of the swing are in sync, in resonance the swing will go higher and higher. The gaps in the rings of Saturn are a good example; the Cassini division or gap is caused by a resonance with the moon Mimas. A rock in orbit in one of these gaps gets pushed

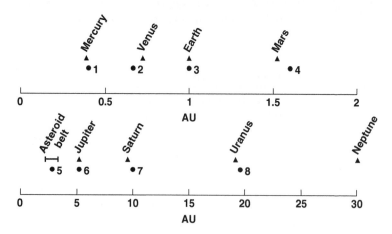

Figure 12.5 The Titus–Bode law. Top: Inner planets. Bottom: Outer planets. The circles are the law's prediction. The labeled triangles are the position of the actual planets.

in the same direction every time Mimas goes by, and soon the gap is cleared. Rocks not in that gap get pushed randomly and so are mixed but not cleared out.

But what of the gap between Mars and Jupiter? This is the region where the asteroid belt is found. We now think there was enough material in this area to form a planet, but because of the gravitational disturbance caused by Jupiter that planet never completely formed. Jupiter would swing by and stir things up before that planet could really congeal. The asteroid belt also gives us another place to look for orbital resonances and it is full of them. The *Kirkwood gaps* are orbits that would be in resonance with Jupiter. There are no asteroids that have an orbital period of $\frac{1}{3}$ that of Jupiter's. There are also none at $\frac{2}{5}$, $\frac{3}{7}$, or $\frac{1}{2}$ Jupiter's period either. So the shape of the asteroid belt is determined by Jupiter.

***

Beyond the asteroid belt are the great Jovian planets, Jupiter, Saturn, Uranus and Neptune. Neptune is the one that really broke the Titus-Bode law. At 30 AU, it was too close and just did not fit. These planets have swept up most of the stray hydrogen and helium in the solar system. In fact, there is so much mass in Jupiter that it is close to becoming a star.

But the Jovian planets are not the end of the solar system. We all know that out there in the cold dark hinterlands of the system lies Pluto, formally the ninth planet.

I find it curious how many people were disturbed when Pluto was "demoted" from a planet to a *dwarf planet*. It seemed as if this mysterious, distant, ice ball was being "relegated," especially when it had not done anything wrong. It just continued to be what it always was. True, it had a tilted orbit that was also so eccentric that it was sometimes closer to us than Neptune. But that was not really the problem. The problem was that Pluto was not really unique. In 2005 Eris was discovered. It is bigger than Pluto but we just had not seen it before because it is much farther away. It became apparent at that time that we were soon going to have a long list of new planets if we accepted everything. In fact Eris was named after the goddess of discord, because it was recognized that its discovery would cause a disruption in the astronomical community.

The *International Astronomical Union* (IAU), which is the scientific body that has jurisdiction over the naming of these bodies, had a long and difficult debate about what is a planet and whether Pluto was to be included. In the end they defined a planet as:

1. It orbits the Sun.
2. It has sufficient mass to force itself to be spherical.
3. It has "cleared the neighborhood" around its orbit of other material.

It is this last criteria that Pluto, Eris and even Ceres, a body found in the asteroid belt, have failed on. Clearing the neighborhood means that it is the dominant body of the region. It may accumulate material, acquire satellites or rings, or eject material. But it must be big enough and have resided in an orbit long enough to be the dominant object of the orbit.

Ceres, Pluto, Haumea, Makemake and Eris are now classified as dwarf planets, and there are certain to be others that will get added to this list and to Table 12.3. Pluto, Haumea and Makemake are part of the *Kuiper belt*, which consists of a myriad of small bodies. It is estimated that there are about 70,000 Kuiper belt objects that are greater than a kilometer across. Unlike rocky asteroids, these tend to be frozen chunks of ice: methane, ammonia and water. The Kuiper belt is also the home of Halley's comet and other short-period (less than 200 years) comets. Beyond and around this region is an area called the *scattered disk*, which contains bodies (like Eris) presumably pushed there by the Jovian planets.

And beyond this? As of the time of this writing (2012) the two Voyager spacecraft are at about 120 and 100 AU from the Earth and Sun. They are starting to pass beyond the *heliosphere*, through a boundary called the *heliopause*. The heliopause is where the solar wind meets the interstellar or galactic wind.

Finally we arrive at the *Oort cloud*. This is a vast region of space from a few thousand AU to 50,000 or more AU. It is probably the origin of many comets. One of the things comet watchers have a hard time explaining is why we continue to see them. Comets are as old as the solar system (4–5 billion years) and most comets make only a few passes near the Sun before they collide with either the Sun or another planet. Shoemaker-Levy 9 was a comet that, in 1994, gave us a spectacular reminder of this when it broke up and collided with Jupiter. By now we should expect all comets to have collided with something. But new comets arrive every year.

Table 12.3 Some properties of the dwarf planets. We should expect more dwarf planets to be added.

Dwarf planet		Distance from Sun ($\times 10^{12}$ m)	(AU)	Radius (km)	Mass ($\times 10^{21}$ kg)	Period (years)	Discovered (year)
Ceres		0.41	2.77	490	0.94	4.60	1801
Pluto	♇	5.91	39.48	1150	13.05	248	1930
Haumea		6.45	43.13	650	4.0	283	2004
Makemake		6.85	45.79	710	~3	310	2005
Eris		10.12	67.67	1160	16.7	557	2005

The Oort cloud was postulated as a source of comets. The cloud is made of frozen chunks of ice out beyond the Kuiper belt and the scattered disk. The chunks of ice orbit slowly out there in darkness for eons until disturbed from their orbits by something, perhaps a passing star or a chance encounter with another Oort cloud object. The ice ball may then be sent into the inner regions of the solar system were we see them. In fact in the last few years astronomers have started directly seeing these distant objects.

***

50,000 AU (7.5×10^{15} m) is only a quarter of the distance to the nearest star. However, on our logarithmic scale we have traveled a tremendous distance. Starting at the radius of the Earth (6.4×10^6 m) we have traversed over nine orders of magnitude. Most importantly, at this scale, we have seen gravity take over as the primary shaping force of nature. It makes planets spherical, it creates rings, it even divides ring with resonances. This weakest of forces happens to have a very long reach.

One way to think about 7.5×10^{15} m is to think about light traveling that distance. Light was too fast for Galileo to measure with his lanterns, but we have now measured its speed at 3×10^8 m/s. Light can travel around the Earth seven times in a second. It takes light just over a second to reach the Moon and about $16\frac{1}{2}$ min to cross the Earth's orbit. It takes light about 4 hours to travel from the Sun to Neptune. Radio signals from either of the Voyager spacecraft take over half a day to get to the Earth. Finally, it takes light almost a year to reach the outer limits of the Oort cloud, the edge of our solar system as we understand it.

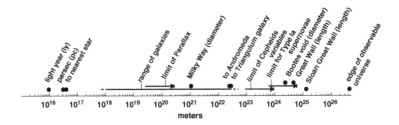

❧ 13 ❧

From the Stars to the Edge of the Universe

In China they call it the *Silver River*, whereas in Japan they trade the description for its location and call it the *River of Heaven*. When you look into the night sky you can easily imagine that that streak of silver-white that arcs across the inky void is a river. It has coves and narrows, dark islands and an undulating shoreline. In Ukraine it is called the *Way of Chumak*. The Chumak were salt traders and one can imagine grains of salt spilled across the sky. In Sweden it is the *Vintergatan*, the winter street. Not only is it best viewed in the winter at northern latitudes, but it does look like moonlight on a road covered with new snow.

In most of Europe it is called the *Milky Way*, a translation from the Greek word, $\gamma\alpha\lambda\alpha\xi\iota\alpha\zeta$ or *galaxias*, which also gave us our word galaxy. The name is traced to the mythology of ancient Greece. Zeus wanted the infant Hercules, his half mortal son, to be nursed by his divine wife Hera and so set the baby on the sleeping Hera. However, when Hera awoke and found a strange baby nursing she pushed Hercules away and milk was spilt across the sky.

Not all Greeks were storytellers and poets. Democritus (450–370 BC) was the philosopher who looked at solid matter and imagined that it could be build up of a myriad of atoms. He also looked at the night sky and proposed that the Milky Way is also a myriad, this time of distant stars.

***

In this chapter we will look at the remaining scales of nature, out to the edge of the observable universe. We will talk about why things have the shapes and sizes they do, as well as how we know this. It is one thing to say that the Crab nebula is 11 light years (or 10^{17} m) across or the breadth of the Capricornus void is 230 Mly (2×10^{24} m). It is far harder to measure that distance. We cannot simply pace it off or send a probe.

In the last chapter we traveled a long way in our solar system; the Oort cloud is $\sim 7 \times 10^{15}$ m out there. But the universe is full of a lot of things we have not yet touched on: nebulae, black holes, galaxies, galactic groups and clusters, even cosmic walls and voids. When you step out at night and look up all those things are there. But what you primarily notice are the stars, including that collection that we call the Milky Way.

***

In all of our discussion of the solar system in the last chapter we left out the largest object, the Sun. Stars come in a great range and variety. There are red giants, white and brown dwarfs, ancient stars and stars that burn bright. Our own star is in the middle of these ranges. It is very average, which is convenient since when astronomers describe stars they do it by comparing it to our own. When an astronomer measures the size of Altair they may report it as $1.8M_\odot$, with a radius of $2R_\odot$. The symbol $\odot$ for sun is part of the same system of symbols that uses $\oplus$ for Earth, $\female$ for Venus and $\male$ for Mars. So we could write the mass of our Sun as: $m_\odot = 333,000 M_\oplus = 1.989 \times 10^{30}$ kg.

The Sun contains 99.96% of the mass of the solar system. This is so similar to the fact that a proton contains 99.5% of the mass of a hydrogen atom that it is worth looking into the parallels. The radius of the Sun is $R_\odot$ or 6.96×10^8 m. The radius of the orbits of the outer-m~~ ~t Neptune is about 30 AU or 4.5×10^{12} m, so there is a bit ur orders of magnitude between them. That is only a tenth between the size of the proton and the orbit of the elec- 'r, Neptune is not really the end of the solar system; the d the heliopause are much farther out and closer to the hydrogen atom ratio, so perhaps the analogy is not so bad. Of course if we include the Oort cloud, which is six or seven orders of magnitude bigger than the Sun, the analogy also collapses.

How do we understand the size of the Sun? It has a diameter of 1.4×10^9 m, almost a million miles. It is about 100 times the diameter of the Earth and about twice the diameter of the Moon's orbit. It takes between 4 and 5 s for light to travel this far. A truck or bus might drive this distance in their lifetime, but very few cars will make it. Airline jets will fly this distance in a little less than a year of service. Did that transatlantic flight of five hours seem long? It would take about 60 days for a jet to fly 1.4×10^9 m, or six months to fly around the Sun.

The Sun is an average star, neither the smallest nor the largest, but instead someplace in the middle of the range of star sizes. That range is due to a balance between nuclear forces and gravitational forces. Briefly, a star is formed when a cloud of gas in space collapses in on itself because of the gravitational attraction of every atom and molecule in that cloud. As it becomes denser, the probability of the nuclei of those atoms colliding increases and with collisions comes nuclear fusion and the release of energy in the form of heat and light. If the amount of matter is too low it could form a planet instead, with the atoms keeping their separation distance as everyday matter does on Earth. The lower limit is called the *hydrogen burning limit* and is about $0.08 M_\odot$.

At the upper limit, with too much matter a star would burn extremely hot and bright, and in that fury blow away excess material until it reached a more stable size. This balance between the outward radiation pressure and the inward gravitational pressure is called the *Eddington luminosity* or the *Eddington limit* and predicts that stars will not exceed about $150 M_\odot$.

Observations back this up. The smallest star measured is GLE-TR-122b (with billions of stars to catalogue, some have less-than inspiring names). It has a mass of $0.09 M_\odot$, just over the hydrogen burning limit. Actually there are smaller stars out there, but they are not burning purely hydrogen and so have smaller theoretical limits.

At the other end of the size scale, large stars burn fast and furious. A star with a mass of $150 M_\odot$ will last only a few hundred million years, a lifetime that is perhaps only 5% of that of our Sun.

For a long time the biggest star we knew of was *Eta Carinae*, or η *Carinae*, which gives us a reason to look at the way the brighter stars are named. The name η Carinae tells us that this is the seventh brightest star in the constellation Carinae. In fact we should be able to go to Ptolemy's star catalogue, the *Almagest*, and find it listed with that name. Except you don not. In the *Almagest* you will find it listed in the constellation Argo. This is because long after Ptolemy's time it was decided that the constellation Argo was too big, and it was subdivided, leaving the unchanging sky . . . changed.

The more people have studied η Carinae the more curious it is. It is believed to have had a mass of about $150M_\odot$, near the Eddington limit, but to have lost about $30M_\odot$ over time. It is embedded in a nebula, which makes it hard to observe, and it has a smaller companion star. But as we have been able to look out farther and farther we have found more stars near the Eddington limit. HD 269810 in the Large Magellanic Cloud, an area outside the Milky Way is about $150M_\odot$.

But the word "biggest" implies a size a not a mass. For a long time VY Canis Majoris was the biggest star that had been measured, with a radius of about $1,400R_\odot$ or 6.6 AU. A recent analysis has lowered the measurement by about 15% and there are new candidates vying for that title of biggest. Still VY Canis Majoris has a radius slightly bigger than the orbit of Jupiter. Despite its size, it falls well below the Eddington limit, with a mass of only $17 \pm 8M_\odot$.

But just when we think we understand stars we find an exception. R136a1 was characterized in 2010 and found to have a mass of about $265M_\odot$, which is well over the Eddington limit. In fact it is thought that it had substantially more mass, but has shed over $50M_\odot$ over the last million years, as it works its way down to a more stable mass. Because the theories of star formation are so well founded, and they say that stars cannot form over $150M_\odot$, it is believed that stars like R136a1 were actually created out of the merger of multiple stars.

***

One of the first attempts to measure the distance to a star was made by Christiaan Huygens (1629–1695). Huygens was a Dutch astronomer, a contemporary of Newton and Cassini, and an early supporter of Rømer's measurements. He made a series of tiny holes in a screen and let sunlight shine through them. He then picked out the hole that he estimated let through the same amount of sunlight in the day as what he saw from Sirius, the *Dog Star*, at night. He then calculated that the angular size of his hole was 1/30,000 the size of the Sun and so Sirius was about 30,000 AU away. It was an ambitious attempt and Huygens underestimated the distance by a factor of about 20, which was due to the fact that Sirius is about 20 times more luminous than our Sun, but he had a rough estimate with which to start.

The best method to measure the distance to stars is *stellar parallax*. We have already encountered solar parallax when measuring the distance to the Sun. In that case an astronomer would measure the position in the sky of the Sun from two locations widely separated, ideally from opposite sides of the Earth, to get the longest baseline. In stellar parallax an astronomer is surveying the distance to the star by constructing a triangle. One measures the position of a star in the sky and then repeats that measurement about six months later when the Earth has moved halfway around its orbit (see Figure 13.1). One side of the triangle is 2 AU or 3×10^9 m, the diameter of the Earth's orbit. But this is a small number, a short distance when compared to the distance to the stars. These triangles are incredibly long and thin. Using Huygen's estimate, the height is 15,000 times the length of the base. In the end it is not really the angle that matters, but how that angle changes, and that change must be measured from our moving, spinning Earth.

Stellar parallax is actually an old idea. Tycho Brahe (1546–1601) did not accept Copernicus's thesis based on the lack of observed stellar parallax and long before him Archimedes, in *The Sand Reckoner*, based his size

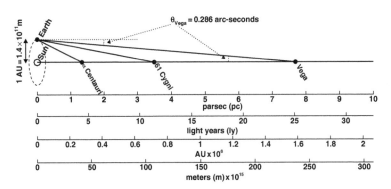

Figure 13.1 Parallax of a near star. Note that the horizontal and vertical directions are not to scale so the angles drawn here are about a million times too large.

of the universe upon the unobserved parallax. Edmond Halley, about the year 1700, looked for the effect by carefully measuring star positions. What he found was that his measurements did not match ancient catalogs because stars move over time. We now call this *proper motion*. Stars are not rigidly fixed on the spiral arms of the galaxy, they chart their own paths across space. James Bradley (1693–1762), the third Astronomer Royal again tried to measure parallax by raising the standards of precision angular measurements and discovered an effect called the *aberration of light*. As we orbit the Sun, sometimes we are moving towards a star and so along the path of the light ray, and sometimes we are moving across the path. Our motion and the finite speed of light will combine to make stars appear at different angles. Bradley did not see stellar parallax, but he did measure the speed of light.

By the 1830s measuring parallax and the distance to another star had become an obsession, a holy quest, for the international astronomical community. Finally, in 1838 three astronomers published successful measurements. First, Friedrich Bessel (1784–1846) measured the parallax of 61 Cygni at 0.314 arc-seconds. Then Friedrich Georg Wilhelm von Struve (1793–1864) reported the parallax of Vega. Finally Thomas Henderson (1798–1844) reported just under an arc-second for the parallax of Alpha Centauri. There are 60 arc-minutes in a degree and 60 arc-seconds in a arc-minute. So 0.314 arc-seconds is one ten thousandth of a degree. That is like a 2-mm shift as viewed from a kilometer, or a 2 hair-width shift as viewed from 100 m. A good telescope can easily see a hair at 100 m, but it is harder to look at it six months later and see the shift. Also one has to measure the proper motion and the aberration of light and remove those effects.

Measuring parallax was a laborious task. So the first trick was to pick a few stars that were promising candidates, stars that you thought might be near. Astronomers reasoned that nearer stars would most likely have greater proper motion, much like a person walking close by you will change their angular position faster than a distant walker. They would then chart these stars over time compared to the more distant stars behind them. Over a few years the candidates would march across the star field, two steps forward and one step back, much like the retrograde motion of the planets that inspired the complex epicycle theory of Ptolomy as well as Copernicus's heliocentric model. The step back was an annual effect caused by the Earth's motion around the sun, or parallax.

The distance unit of choice when measuring parallax
The term parsec comes from the words parallax and arc-
it is defined as the distance away an object is if it shifts
second when the observer has moved 1 AU. So Bessel's meas
0.314 arc-seconds translates to 1/0.314 or about 3 parsecs (about 10^{17} m).
Henderson picked out as his candidate the star that is the nearest one
you can see without a telescope: Alpha Centauri. Modern measurements put this at 0.747 arc-seconds, which is 1.3 parsecs, 4.3 light years
or 4×10^{16} m (see Table 13.1). If a Voyager probe was headed in that
direction it would take about 75,000 years for it to reach the star.

There is one other unit of measurement that is used especially in
popular literature: the lightyear. It is a unit whose very name invokes
vast distances, the deep dark void of space, and the lonely stellar beacons
of light. A lightyear (ly) is the distance light travels in one year. A year
contains 31 million seconds, so a light year is about 10^{16} m. Whereas a
lightyear is descriptive and has caught the popular imagination, professional astronomers tend to favor the parsec, a unit that is directly tied
to the measurement technique.

From the surface of the Earth stars tend to shimmer and twinkle
due to the atmosphere, which may be a pretty effect, but does not
help when trying to make precision measurements of their positions.
Ground-based measurements are limited to about 20 to 50 parsecs. With
stars in our region of the galaxy spaced about every parsec, that means
we can measure the distance to tens of thousands of stars, which is a

Table 13.1 List of nearby stars.

#	Name	Distance lightyears	Apparent magnitude	Parallax milliarcsec
1	Sun	—	−26.74	
2	Proxima Centauri	4.2421	11.09	768
3	α Centauri A	4.3650	0.01	747
4	α Centauri B	4.3650	1.34	747
5	Barnard's star	5.9630	9.53	547
21	61 Cygni	11.403	5.21	286
	Vega	25	0.03	130

Note: Proxima Centauri is the closest star, but with a apparent magnitude of 11.09 it cannot be seen without a telescope.

drop in the bucket compared to the whole galaxy. So for four years, from 1989–93, the Hipparcos space telescope orbited above the atmosphere and measured stellar parallax out to 100 to 200 parsecs, which means the distance to about a million stars. Finally, over the next few years the GAIA mission should extend that measurement out to as much as 8,000 parsecs. With these types of measurements there is not a firm cut-off beyond which you cannot measure. It is just that at long distances the results have great uncertainty. For example, GAIA plans to look at stars near the galactic center, which is about 10,000 parsecs away, but at that distance it will expect an accuracy of only 20%. That is 10^{17} m, which means we have a long way to go to measure the distance to the edge of the observable universe.

We cannot directly measure the distance to a star beyond a few hundred (and soon few thousand) parsecs. But this is similar to the limit optical microscopes have; they cannot see atoms directly. Likewise we could not see quarks directly. But we can use other techniques to push back the limits of what we can measure. To find a new technique we can also ask what it is besides distance we can measure in these nearby stars. Two things stand out: we can measure the brightness of a star and its color.

How bright a star is is called its magnitude, as was mentioned in Chapter 4. What we directly measure is its *apparent magnitude*. But how bright a star appears depends upon how bright it really is, as well as how far away it is; that is, the amount of light we see drops with increasing distance because the photons from that star are spread over a larger and larger area the farther they radiate out. If we know the distance to a star we can calculate how bright it would be if viewed from a standard distance. This is the same thing that was done with the Richter scale. By convention, the standard distance is ten parsecs, and the brightness at that distance is called its *absolute magnitude.*

In about 1910 Ejnar Hertzsprung (1873–1967) in Denmark and Henry Norris Russell (1877–1957) in America, independently recognized a trend in stars that would help measure stellar distances beyond the parallax limit. What they did was take the data of the well-measured stars and plot their color (temperature) versus their absolute magnitude (see Figure 13.2). About 90% of the stars fell on a line that reached from the hotter (blue) bright stars to the cooler (red) dim stars. The trend is striking and we refer to this band as the *main sequence*, and the whole diagram as the *Hertzsprung–Russell diagram*, or sometimes simply as the H–R diagram. This means that if an astronomer measures a star's

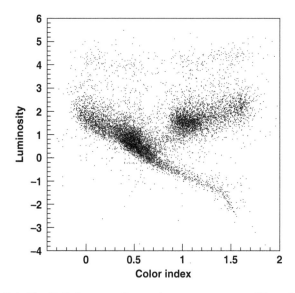

Figure 13.2 The H–R diagram of star color versus intensity. The color index is related to temperature. The luminosity is related to absolute magnitude. The streak of stars from the upper-left to lower-right contains about 90% of the stars and is called the *main sequence*.

color, they know how big and bright it is. Astrophysicists also plot the evolution of stars on H–R diagrams, but for us what is most important is that it correlates color (temperature) to absolute magnitude. So this gives us a new technique for measuring stellar distances. We look at a star and measure its color and apparent magnitude. From its color, using the H–R diagram, we can figure out its absolute magnitude. Finally by comparing the absolute and apparent magnitudes we can figure out how far away it is.

This technique is called *spectral parallax*, even though it does not really use parallax at all. It is sometimes also referred to as the *third rung of the cosmic ladder*. The first rung was determining the AU, now best done with radar bouncing off planets. The second rung was stellar parallax. Every step outward requires that the previous step be firmly established. It also means that uncertainties in measurements of galactic scales have uncertainties in the AU and the H–R diagram embedded in them.

***

Before pushing our horizon out beyond our galaxy there are two smaller objects to look at: black holes and nebulae. One of the most curious types of objects found by astronomers in the last half century are black holes. These objects have so much mass that light cannot escape from them. They are the ultimate "stealth" object, the astronomical counterpart to quarks. We cannot extract quarks from inside of other particles, but we still know a lot about them. The same is true about black holes. We cannot see them directly, but we can observe their effects upon the things around them, including in particular, and most curiously, light.

Black holes have the ability to bend the path of light, an effect that is sometimes described as *gravitational lensing*. Actually, any body with mass will do this, but black holes do it well and it is an effect we can observe even if we cannot see the black hole itself. Some of the first observational evidence for black holes were double images of stars and galaxies. Light from a galaxy headed in two different directions would pass on either side of a black hole where both light-rays are bent to where we see it. We look in two different directions, on either side of the black hole, and see the same galaxy.

The most common black holes are *stellar black holes*, formed from the gravitational collapse of a big star. After it has burned its fuel and there no longer is a strong radiation pressure pushing the surface out, or in fact keeping the atoms separate, the worn-out star can gravitationally collapse and form a black hole. Stellar black holes typically have masses of about ten solar masses ($10M_\odot$) and radiuses of about 30 km.

There are also *supermassive black holes*, found at the center of most galaxies. We have now observed stars orbiting these black holes. They have masses of the order of $100,000M_\odot$, although their radius may only be about 0.001 AU, which is about half the distance to the moon. Even these supermassive black holes really are not very big.

***

The other type of object in our galaxy are nebulae. In some sense, nebulae might not even make our list of objects since they really do not have a structure that is determined by a force, in the way that gravity forms planets and the electromagnetic force forms atoms. But they are very big objects that play an important role in star formation and are also markers of the demise of a star. So we will briefly look at them.

One of the most studied nebulae is the Crab nebula, having "historic significance." This nebula is the remnant of a supernova, a star that died in a great explosion, spewing out its remaining gases and newly formed heavy elements. With an apparent magnitude today of about +8, we cannot see it with the naked eye; our eyes can see celestial objects with magnitudes down to about +6. But it does not take much of a telescope to see it; even a good pair of binoculars will resolve it. It was first seen as a nebula by John Bevis in 1731. In the early part of the twentieth century astronomers had taken several photographs of it and realized that over the years it was expanding. By measuring the expansion rate they could work back to when it had been just a point, and arrived at its birth year of about 1050.

Historic records show that in July of 1054 a new "star" was seen in the sky in the constellation of Taurus, at the tip of the bull's horn. That is the same location as the Crab nebula. It was so bright that it could be seen in daylight for the first three weeks and at night for the next two years. It was recorded in both Europe and China, where it was called a guest "star," and upset a lot of ideas about the permanence of the celestial sphere.

By looking at the stars in front and behind it, we have now measured the distance to Crab nebula at about 6,500 ly and its size as 11 ly across. It continues to expand at about 1500 km/s. Since we on Earth saw it explode in 1054, it really exploded in about 5,450 BC, because light takes a while to get here.

Supernovae are the only source of heavy elements; that is, elements heavier then iron. They are created via nucleosynthesis in that moment of destruction. But nebulae have a more active role at the other end of the stellar lifetime. Nebulae, with an abundance of gasses, are sometimes called the nurseries or incubators of stars. This is where an embryo star can sweep up the raw materials to form itself and be born.

Another famous nebula is the Orion nebula. With an apparent magnitude of +3, it is visible to the unaided eye, located in the dagger hanging from Orion's belt. This cloud of gas glows because of the stars that reside within it. Presently astronomers have identified over 700 proto-stars and 150 proto-planets with in it. It is about 24 ly (2.3×10^{17} m) across and 1,340 ly (1.2×10^{19} m) away.

Location, location, location: one of the largest nebulae in the sky is the Great Nebula of Cerina. It is an order of magnitude larger than the well-studied Orion nebula and even contains the star η Carinae, the

massive star we talked about earlier in this chapter. But it is not well
known or studied because it lies in the southern celestial hemisphere
and is not visible from Europe, Asia, or North America, where most of
the telescopes and astronomers are.

***

But we need to move on. We still have seven orders of magnitude before
we reach the edge of the observable universe.

The next biggest objects are galaxies. Our galaxy, the Milky Way, is a
pretty big one. For many years we assumed that the Milky Way looked
very much like the Andromeda galaxy. If that were the case it would
simplify the study of our home galaxy, since we have a good view of
Andromeda. Our own galaxy is hard to see from the inside. It is like
standing in the midst of the crowd in Times Square on New Year's Eve
and trying to figure out how many people there are and where the
edge of the crowd is. With visible light we can only see about 3 kpc
(kpc = kiloparsec) (10,000 ly = 10^{20} m) through the Milky Way because of all the dust and gasses. We expected, because of looking at
Andromeda, that this was only about 10% of our galaxy. In fact, it was
only with the development of radio astronomy that we were able to
see farther and figure out in which direction was the galaxy center (see
Figure 13.3).

We now think the Milky Way is a *barred spiral galaxy*, which means that
in the center is a *bar*, an elongated region thick with stars. Extending
out from the ends of the bar are spirals, much like what we see in
Andromeda. Our solar system is located in the *Orion spur*, a small finger
that sticks out of the *Perseus arm*, one of the major spirals of the galaxy.
When naming regions of space we see the same classic names again and
again. What a name like the Perseus arm tells us is that this arm is in the
same direction of the sky as the constellation Perseus. The same scheme
named the Andromeda galaxy, the Sagittarius arm and so forth.

The whole disk of the Milky Way, bar and arms, is about 100,000–
120,000 ly (10^{21} m) across and relatively thin, perhaps only 1,000 ly. The
bar itself is a few times thicker. (This is one of those confusing cases
where the size of the Milky Way is most often quoted in light-years, but
most other things in parsecs. I think it is because the number 100,000 is
easier to remember than 30,000 parsec, or 30 kpc.) This bar is an interesting feature, and we now know that about two-thirds of all spiral galaxies

Figure 13.3 A map of the Milky Way. The Milky Way is about 100,000–120,000 ly across, or 10^{21} m. Courtesy NASA/JPL-Caltech/R. Hurt (SSC).

have them. Recent computer simulations of galactic evolution indicate that they are a normal part of a galaxy's life cycle; they form, dissipate, and reform over time. They are a normal mode of the galaxy, much like atoms have wavefunctions with different shapes. Also, when a bar is present, that tends to sweep up gasses and helps in star formation.

But there is more to our galaxy than just the disk. Surrounding the galaxy and gravitationally tied to it are *globular clusters*, collections of older stars. These clusters are not confined to the plane of the galactic disk and instead mark out a spherical region about 100,000 ly in radius; that is, as much as twice the radius of the disk.

A disk of stars swirling in space, rotating every hundred million years or so: it is a simple picture of our galaxy, and most other galaxies, except it has a problem. It does not explain the structure we see. The rate of rotation should follow Kepler's laws and the arms should get mixed together. In fact it is this problem that has led to the hypothesis of dark matter. There must be more mass out there beyond the galaxy's halo to give it the structure we observe.

***

A lot of what we know about galaxies actually comes from seeing other galaxies. After all, we live in just one of the several billion examples we can see. But for many years we did not even recognize that galaxies were out there, because they just appeared as distant, fuzzy, nebula-like objects. They were too far away to measure with stellar parallax and too fuzzy for us to pick out main sequence stars. It is true that some people viewed them as separate objects in space. As far back as 1755, Immanuel Kant introduced the term "island universe," but without any real support. What was needed was a measurement of the distance to these nebulas, or island universes. We needed the next rung of the cosmic ladder.

Back in Chapter 4 we described the star magnitude scale introduced in ancient times and then precisely defined by Pogson in 1853. With the introduction of astrophotography, the method of comparing star brightness became even more objective. You could now compare images of stars side by side on photographic plates. You could even compare images of the same star taken at various times, which led to the surprising discovery that some stars had changing magnitudes. Their brightness waxed and waned like the tides.

Henrietta Swan Leavitt (1868–1921) worked at the Harvard College Observatory, where she measured and recorded the brightness of stars from photographic plates. People before her had understood that some stars vary in brightness, in particular a type of star called a *Cepheid variable*. Cepheid variables are named after the star Delta Cephei found in the constellation Cepheus and are notable because their brightness rises and falls with great regularity. What Leavitt realized was that the period of change was related to the absolute brightness of these stars.

Leavitt was looking at stars in the Magellanic Cloud, a cluster of stars all at about the same distance from the Earth. She could determine the absolute brightness and therefore their distance from Earth. Her discovery gives us our next rung. With Cepheid variables we can measure stars as far away as tens of millions of parsecs. The most distant object measured by this method is a star in the NGC 3370 galaxy in Leo. Its distance from Earth has been measured at 29 Mpc (10^{24} m).

In fact it was Cepheid variables that resolved one of the great scientific debates of the early twentieth century. Were things like the Andromeda galaxy a "nebula" or separate "island universe"? Edwin Powell Hubble (1889–1953) used Cepheid variables in the early 1920s to measure the distance to neighboring galaxies and concluded that

Figure 13.4 The Andromeda galaxy. Andromeda is about 2.5 Mly (2.4 × 10^{22} m). The well-illuminated part is about 100,000 ly wide, although it may in fact be a few times wider. Courtesy of NASA.

these were indeed distant objects, island universes, separate and distinct from our own galaxy. Andromeda is about 2.54 Mly (2.4 × 10^{22} m) away and the Triangulum galaxy just a bit beyond that. That means they are about 25 times farther away than the width of our own galaxy. Andromeda subtends a wider angle than the moon; it is just very faint (see Figure 13.4).

Cepheid variables will help us measure distances out to twenty or thirty times farther than Triangulum galaxy, but first let us look at the world within a few million lightyears of Earth. This region of space is called the *local group* and consists of three major galaxies (Andromeda, Triangulum and the Milky Way galaxies) as well as about four dozen *dwarf galaxies*. All of the galaxies in the local group exhibit gravitational cohesion, with many of the dwarf galaxies identified as satellites of one of the major galaxies. This local collection gives astronomers about 700 billion or more stars to look at. With stars living tens of billions of years, there should be several stars being born and dying every year.

***

To continue to look deeper in space, and know how far we are looking, we need new techniques. Techniques such as using the main sequence or Cepheid variables are collectively referred to as *candles*. A candle is some feature of a star or galaxy that will tell us its absolute brightness or magnitude. Then, by comparing this figure to the object's apparent magnitude we can figure out how far away it is from us. There are a number of candle techniques, but we will only add two more: the Tully–Fisher relationship and supernovae.

The Tully–Fisher relationship is a correlation between the brightness of a galaxy and its rotation rate. The trick here is to measure the rotation rate. We cannot just take a photograph, wait a few years and take a second one and see how things have changed. Our own galaxy takes about 200,000,000 years to rotate, so we would have to wait a very long time to see a noticeable change. But the stars are actually moving quickly; our solar system moves at 230 km/s around the disk. So instead of looking at the rotation directly we will look for the *Doppler effect* of motion.

***

The Doppler effect is what happens when a train blasts its horn as it passes by. It sounds as if the horn has changed from a high pitch as it approaches to a low pitch after it passes. The motion of the source of sound affects the way we hear it, be it a siren, a race car, or a jet plane. If the source of light is moving fast, it will affect the way we see it. The effect is caused by the source either chasing after the waves it made, or running away from them. If a fishing bobber is bobbing up and down in the middle of a pond the water waves will spread out evenly in all directions around it (see Figure 13.5). However, if the bobber is moving to the east, because the angler is reeling it in or a fish is dragging it, the bobber will chase the eastern waves and leave behind the western ones. An observer on the eastern shore will see short, frequent waves, whereas the western observer will see longer wavelengths.

If instead of a bobber on a pond, the source was a yellow light, as it moved towards us the waves would bunch up, the frequency would be higher and the wavelength shorter. The yellow light would appear to shift towards the blue end of the spectrum. If, however, the source was moving away from us it would red-shift. For most things the effect is very small because the waves are moving so much faster than

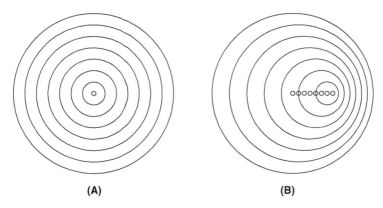

(A)   (B)

Figure 13.5 The Doppler effect in water waves. (A) A stationary bobber in a pond radiates concentric waves. (B) A bobber in motion chases its waves. An observer on the right will see short, frequent waves, on the left longer waves.

the sources. Our solar system, traveling at 230 km/s around the center of the galaxy, is moving at 0.07% of the speed of light, a small but measurable effect.

The way that Tully–Fisher use this information is to look at a well-known spectrum line that is found in the light from a distant galaxy. Ideally you would like to look at one edge of the galaxy and see blue-shift as it spins towards you and on the other edge red-shift as it spins away from you. But distant galaxies can appear point-like, in which case the shifts merge together; instead of separate blue- and red-shifts you just see a broadening of the line. But that still works. The broader the line, the faster the rotation, the brighter the galaxy. The Tully–Fisher method has been used to measure distances to galaxies 200 Mps $(6 \times 10^{24}$ m) away.

***

There is one last candle technique we will describe: supernovae. Remember that the way a star works is that is balances the outward radiation pressure against the inward gravitational pressure. Near the end of its life it may just fizzle and collapse inward. However, in some cases, as it cools and collapses there may be enough pressure such that it can burn some of the heavier elements in it, for example carbon. "Burn" is really an understatement: it ignites—cataclysmically—into a massive explosion. The explosion may only take minutes, with the results fading in days (like the Crab nebula), but during that time the star may radiate as much energy as our Sun does in its whole lifetime. So much energy means that a supernova can briefly outshine its host galaxy!

There is one special type of supernova, known as the *Type Ia supernova*, that is of particular interest. The star must be of just the right size for this type of supernova to happen, which means it has a unique absolute magnitude. All Type Ia supernovae give up about the same amount of light and energy. They are also very bright, which makes them ideal candles. The only problem is that they are brief and uncommon. Fortunately there are a lot of candidates out there. In fact these supernovae are so bright that we can see and measure them out to about 1 Gpc (gigaparsec; 3×10^{25} m). With these distance scales we can map a vast section of space, well beyond the local group of galaxies.

***

Space is filled with *galactic groups* and *galactic clusters*. These are clumps of galaxies that are gravitationally bound. Groups are generally smaller than clusters, but at this time there is no hard definition. Galactic groups generally have 50 or fewer galaxies and are only 1 to 2 Mpc across. Nearby Earth is the Canes Venatici I group, about 4 Mpc away, as well as several different *Leo* groups. The Leo groups are located beyond the constellation of Leo, but all at different distances.

Larger than the groups are clusters, like the Fornax cluster with almost 60 galaxies, located 19 Mpc (6×10^{23} m) away and the Eridanus cluster with more than 70 galaxies and found 23 Mpc (7×10^{23} m) away. But what dominates our part of the universe is the Virgo cluster. The Virgo cluster contains over 13,000 galaxies. Since it is about 5 Mps across, that means galaxies are spaced about every 100,000 pc, which we can compare to the diameter of our own galaxy—about 30,000 pc across—just to get an idea of the density of galaxies in a group or cluster.

Moving outward an order of magnitude we see that the local group and the Virgo cluster are part of a bigger collection of objects, which is called the *Virgo supercluster*, named after the dominant cluster in the region. This supercluster contains at least 100 galaxy groups or clusters

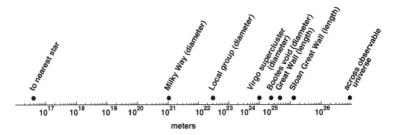

Figure 13.6 Comparison of the size of galaxies, group, superclusters and the universe.

***

and is about 33 Mpc (10^{24} m) across. We are now approaching 1% of the diameter of the universe (see Figure 13.6).

***

To measure the remaining 99% of the universe we must return to the work of Edwin Hubble. As we saw a few pages ago, in the 1920s Hubble used Cepheid variables to measure the distance to the Andromeda and Triangulum galaxies. But he did not stop there. He went on to measure the distance to a total of 46 galaxies and discovered a relationship between the distance to a galaxy and its red-shift. In 1929, only four years after showing that Andromeda and Triangulum were beyond the Milky Way, he published the correlation between red-shift and distance. We now call this *Hubble's law*. The farther away a galaxy is, the faster it is moving away from us. A few near by galaxies are blue-shifted; that is, they are moving towards us. Most famously, Andromeda is on a collision course with us and is expected to collide in four billion years' time, but this is just local motion. The trend for the vast majority of galaxies is that they are moving outward. This is solid evidence for the Big Bang and the expansion of the universe.

Einstein's theory of general relativity, the most fundamental theory we have of gravity and space, makes a lot more sense in an expanding universe than in a static or steady-state universe. But in the early decades of the twentieth century the overwhelming opinion of astronomers was that space is the same as it has always been and will be. This prejudice was so strong that even Einstein accepted it. Einstein's original formulation of general relativity predicted an expanding universe, but he added a term to his equations to cancel out the expansion. Later he would call this his "greatest blunder." In fact in January 1931, he visited Hubble at the Mount Wilson observatory to thank him for finding the expanding universe and setting cosmology in the right direction.

The fact that distant galaxies are moving away from us faster than nearby galaxies, and so have a greater red-shift, can be used as a new technique for measuring vast distances. Hubble's law has now been verified over thousands of galaxies, the distances of which we can independently measure with techniques such as Cepheid variables, Tully–Fisher, supernovae, as well as other methods. We have arrived at our last rung of the cosmic distance ladder. Hubble's law states:

$$v = H_0 D \quad or \quad D = v/H_0$$

$$H_0 = 74.3 \pm 2.1 \frac{\text{km/s}}{\text{Mpc}}$$

or the velocity (v) of a galaxy is Hubble's constant (H_0) times the distance to that galaxy (D). So, for example, if we observe a distant galaxy with a velocity of 7000 km/s (2% of the speed of light) then that galaxy is 100 Mpc (3^{24} m) away. This is a fantastic tool. Now all an astronomer needs to do to measure the distance to a galaxy is to identify a line in the spectrum from that galaxy, perhaps a transition in hydrogen, and see how it has shifted compared to a non-moving source due to the Doppler effect. From that you know how far away the galaxy is. Now we can build a three-dimensional map of the universe with millions of galactic clusters, out to billions of parsecs.

What we see is called the *large-scale structure of the universe*, and it is beautiful and astonishing. When we plot the position of galaxies in three dimensions, not only do we find galaxies in groups, clusters and superclusters, but these collections of galaxies line up to form wispy filaments that reach across deep space (see Figure 13.7). The area of the universe with stuff in it seems to be gathered into sheets and lines. The areas with nothing in them are bubbles of inky black void. This structure has been described as *foam*, ironically the same word that has been used to describe the world at the Planck length. But to me what makes these structures so intriguing is that maps of the universe have names sprinkled across them. Perhaps the most famous is *the Great Wall*,

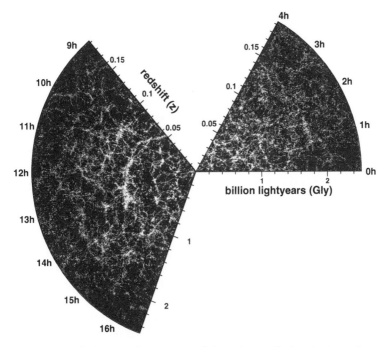

Figure 13.7 The large-scale structure of the universe. Each point is a galaxy. Galaxies tend to bunch together into walls, and leave bubbles or voids. Right and left on this diagram are looking perpendicular to the disk of the Milky Way. We cannot see very far in the up and down direction because of the dust in gasses in our own galaxy. This data is from the Sloan Digital Sky Survey. Courtesy of Michael Blanton and the Sloan Digital Sky Survey (SDSS) Collaboration, http://www.sdss.org.

a structure 60 Mpc away, 150 Mpc long, 90 Mpc wide, but only 5 Mpc thick.

Areas without galaxies are also named, for example the *Boötes void* and the *Capricornus void*. It is odd to think of us humans, not even 2 m tall, naming regions of such magnitude. There is a bubble or void beyond the constellation Eridanus, which would take light about a billion years to cross. Originally it was named the *WMAP cold spot*, after the satellite that spotted it. This is like Beagle Bay in Australia being named after the ship that dropped anchor there and charted it. The *Sloan Great Wall* is named after the Sloan Digital Sky Survey, the project that charted it. This wall, at 420 Mpc (1.3×10^{25} m) long, is the largest known structure in the universe, although some recently spotted large quasar groups may challenge this title. Could Alfred P. Sloan have imagined that his name would be attached to such a vast region when he endowed the Sloan Foundation?

We are not quite at the end of the road. We are talking about structures that are as much as a billion light years across. But as we approach the edge of the universe, things get odd. First off, if you observe something that is a billion light years away you are also looking back in time a billion years because that is how long it took that light to reach us. The most distant thing we can see is the *cosmic microwave background*. These photons have been traveling since shortly after the Big Bang. Cosmologists tell us that for the first 380,000 years after the Big Bang the universe was a plasma and too hot for photons to penetrate. But since then the universe has cooled off and these microwave photons have been traveling towards us. This also forms the edge of what is referred to as the *observable universe*. Light from earlier, and so farther away, cannot be seen.

So how big is the observable universe? This is a complex problem because space and time get mixed together. However, let us start with something simple, like the Hubble constant, $H_0 = 74$ (km/s)/Mpc. It is a curious constant that has distance in both the numerator (km) and the denominator (Mpc), which we can divide to get $H_0 = 74$ (km/s)/Mpc $= 2.4^{-18}$/s$= 1/13.2$ billion years. The Hubble constant is close to 1 over the age of the universe, which is 13.7 billion years. It is tempting to say that the size of the universe is the distance the fastest-moving galaxy has traveled since the Big Bang. Since nothing can move faster than light we might reason that the radius of the universe is about 13.7 billion light years (4.2 Gps or 1.3×10^{26} m). But we would be wrong. The universe is even bigger than that.

Light left that most distant object about 13.7 billion years ago (minus 380,000 years) and traveled at 3×10^8 m/s ever since then to get to our telescopes today. If that photon had an odometer it would read 13.7 Gly (measuring distance while sitting on a photon actually makes no sense). But the space that photon has traversed is now bigger than it was. The universe is expanding. In fact galaxies are not really moving away from us, but rather, the space between galaxies is growing.

The best analogy I know of to help us understand this is to look at the expansion of a rubber balloon as it inflates. If the balloon is covered with spots as it inflates, an ant sitting on any spot will see all other spots move away from it. This analogy is usually used to explain why all galaxies are moving away from us, yet we are not in a unique location. All galaxies see all other galaxies moving way from them.

Now let us take this analogy one step farther. Imagine that our spot-riding ant can walk 10 cm in 10 s. It starts from its home spot and arrives at its destination 10 s later. How far is it back to its home? The ant traveled 10 cm, but its home is farther away than 10 cm since the balloon was expanding during that trip. We see light that has traveled about 13 Gly, but that light shows us objects that are now three times farther away. The radius of the observed universe is about 14 Gps, 46 Gly or 4.3×10^{26} m in any direction.

Finally we can ask "What is the biggest structure in the universe?" The Great Sloan Wall is 420 Mpc across, whereas the whole universe is 28,000 Mpc. Are there any mega-macro-ultra-structures left? Are there collections that are 1000 Mpc across or larger? Apparently not, or at least we do not presently recognize them. This fact is sometimes known as "the end of greatness." If I look at the universe with a resolution of 10^{25} m or 10^{26} m you would see no structure. The universe at that scale appears to be uniform and homogenous.

***

4.3×10^{26} meters from here to the edge of the universe and it continues to grow. In the few seconds it has taken to read this sentence the universe grew by over a billion meters, a million kilometers. The universe grows beneath our feet, like walking on a moving sidewalk in an airport or Alice running with the Red Queen in *Through the Looking Glass,*

"Well, in OUR country," said Alice, still panting a little, "you'd generally get to somewhere else—if you ran very fast for a long time, as we've been doing."

"A slow sort of country!" said the Queen. "Now, HERE, you see, it takes all the running YOU can do, to keep in the same place. If you want to get somewhere else, you must run at least twice as fast as that!"

Through the Looking-Glass
Lewis Carroll

4.3×10^{26} meters. The universe is not only vast, but it continues to grow, and that growth affects light in the sort of ways Alice may have encountered through the looking-glass. And in that twisted way of light and space we may just be glimpsing the end of space—and the beginning of time.

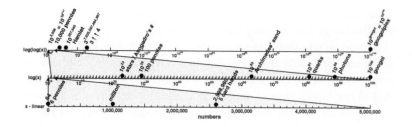

❦ 14 ❦

A Little Chapter about Truly
Big Numbers

We have now talked about the range of time, from the lifetime of particles that decay in a yoctosecond (10^{-24} s) to the age of the universe, 13.7 billion years (4.3×10^{17} s), a span of 41 orders of magnitude. In length scales we have seen a range from the Planck length (1.6×10^{-35} m) to the width of the universe (8.6×10^{26} m), a span of 61 orders of magnitude. But to a mathematician these are small numbers. Even Archimedes was interested in bigger numbers. This book is primarily about the physical universe, but in this chapter I would like to digress once more and see the problems of "How Big is Big" from the viewpoint of numbers.

Archimedes told us how many grains of sands there might be if his ideal universe was filled with sand. I will instead address the question of how many particles there are. That is a big number, but physics needs even larger numbers when we talk about potential combinations, a cornerstone to Boltzmann's statistical mechanics. We will briefly look at combinations as they have to do with entropy, as well as for the pure joy of numbers. And what numbers! They will break our notation scheme and we will have to find new ways of expressing even bigger numbers. Finally we will look at infinity, or rather at a few types of infinity. Some of this will seem like a deluge of abstract mathematics.

Do not let the vastness overwhelm you. Do not treat it as something you have to know, but rather let this mathematics just flow. Let it excite and entertain you.

★★★

As Archimedes explained, it takes 10,000 grains of extremely fine sand to fill the volume of a poppy seed; 40 poppy seeds side by side equal a finger's width and 10,000 fingers' width to a stade. Finally, since the diameter of the universe, according to Archimedes, was no more than 10 trillion stade, the whole volume could be filled with 10^{63} grains of sand. The radius that Archimedes used was 10^{16} m, or between one and two light years. That is a universe whose boundaries would be about half the distance to the next star. The real observable universe has a radius of about 4.3×10^{26} m; that is, ten orders of magnitude larger. If we filled this modern universe with Archimedes' sand it would contain about 10^{95} grains.

Of course the universe is not filled with sand, but what it is filled with is still pretty impressive. The Sloan Digital Sky Survey estimates that the number of stars in the universe is about 6×10^{23}. By coincidence that is about Avogadro's number, the number of carbon atoms in 12 g, or the number of hydrogen atoms in 1 g. However it is *only* a coincidence, because chemist arbitrarily picked 12 g of carbon as a standard mole.

Sand may not fill the universe, but particles do. Amazingly enough, we can calculate the number of quarks in all of space. New deep-sky surveys tell us that the universe is "flat;" that is, it is not curving back on itself. That means that the universe maintains a unique balance between outward expansion, expressed in Hubble's law, and inward gravitational collapse. So we can actually calculate the density of the universe by balancing Hubble's constant and Newton's gravitational constant and obtain the *cosmological critical density*:

$$\rho_c = \frac{3H_0^2}{8\pi G} = 9.3 \times 10^{-27} \text{ kg/cm}^3$$

Since we also know the size of the universe, the mass must be about 3×10^{54} kg. If that is primarily protons and neutrons, that would be 2×10^{81} nucleons (mostly protons) and 6×10^{81} quarks. This is a number that pales compared to the number of grains of sand to fill the universe. It also implies that there are about 15 quarks or 5 atoms for every cubic meter of space. There is a big "if" on these numbers depending whether the world is made up primarily of protons and quarks. Presently we think 5% of the mass of the universe is "ordinary matter," and 95% is dark matter. We do not know what dark matter is and I have just treated it as if it were made of quarks just cobbled together in some unusual way.

We can also look at the number of photons in the universe. The temperature of deep space is $T = 2.72$ K, a few degrees above absolute zero. Back in Chapter 5, we saw how Planck told us how to connect temperature to the number of photons. There are about 411 photons per cubic centimeter or 4×10^8 photons/m^3. That adds up to 1.4×10^{89} photons. These are impressive numbers indeed, but to study the motion of the atoms in a teaspoon of tea, we would need bigger numbers.

***

Back in Chapter 5 we also spent some time trying to understand Boltzmann's description of heat, energy and entropy. We said that a well-ordered system was low in entropy and described the system in terms of six pennies. With six pennies in a line there are $2^6 = 64$ combinations of heads and tails. Of those, only one is all heads. That combination (and the all tails) are the most orderly states. There are 6 ways of having one head, 15 ways of having two heads, and 20 ways to have 50% heads. These 20 arrangements: TTTHHH, THTHTH, TTHHTH, . . . are the most common states and the ones with the greatest disorder or highest entropy. But remember Boltzmann is thinking in terms of the number of atoms in a bottle of air.

If we have 100 pennies, there will be 10^{29} combinations that are 50% heads. If we had ten thousand pennies, the number increases to $10^{3,008}$ combinations that are half heads. That is a number with 3,008 digits! Still, we are no where near Avogadro's number of pennies, but fortunately, in Boltzmann's equation for entropy it is the logarithm of combinations—effectively just the exponent—that matters.

***

Another example of combination is found in a deck of cards. If you are playing cards and are dealt the ace of clubs for your first card, you might think yourself lucky. If the second card is the two of clubs you might feel less lucky. However if you continue to receive the three, four and five of clubs you might be suspicious about the shuffling of the deck and wonder what everyone else was dealt. The dealer may tell you that these things happen. Actually that sequence is just as likely as any other; it just catches the eye. There are 2,598,960 five-card hands you could be dealt. The trick to a game like five-card draw (aside from betting and bluffing) is to have the least likely hand. So a royal flush (4 out of 2 million) beats a pair (1 million out of 2 million). The pair is a high-entropy combination.

Can you shuffle a deck and have all the cards return to their original order? This problem is related to the future of the universe and is called the recurrence time. When you shuffle the deck the cards can take on one of 8×10^{67} arrangements. Given that the age of the universe is 4×10^{17} seconds, it is not an everyday occurrence.

Another famous problem in randomness states that if enough monkeys were to randomly type for long enough they would type out all the books in the British Museum. If ten billion monkeys (a bit more then the human population of the Earth), were to type ten key strokes a second for the age of the universe they would type about 10^{29} characters or 10^{26} pages. Instead of waiting for them to type out the whole contents of the museum, let us only consider Hamlet, which contains about 180,000 characters. If the monkey's keyboard had only 26 letters and a space bar, the probability of randomly typing Hamlet is 1 in $27^{180,000}$ or 1 in $10^{257,645}$. This is a ridiculously large number and so the probability of the monkeys getting it right is incredibly small. $10^{257,645}$ is a number with over 257,000 digits. When we introduced scientific notation with numbers like 3.0×10^{8} m/s we said that we were dropping less significant information and focusing on the 3 and 8 only. Now we are looking at the number of digits only and disregarding what those actual digits are.

One of the classic numbers with a name is the *googol*. A googol is 10^{100}, which is bigger than our count of quarks or photons, but dwarfed in comparison to combinations of pennies or letters. The mathematicians Edward Kasner (1878–1955) and James R. Newman (1907–1966) described the invention of this name in their book, *Mathematics and the Imagination* (1940):

The name "googol" was invented by a child (Dr. Kasner's nine-year-old nephew) who was asked to think up a name for a very big number, namely 1 with one hundred zeros after it. He was very certain that this number was not infinite, and therefore equally certain that it had to have a name. At the same time that he suggested "googol" he gave a name for a still larger number: "googolplex." A googolplex is much larger than a googol, but is still finite, as the inventor of the name was quick to point out. It was first suggested that a googolplex should be 1, followed by writing zeros until you got tired. This is a description of what would actually happen if one actually tried to write a googolplex, but different people get tired at different times and it would never do to have Carnera a better mathematician than Dr. Einstein, simply because he had more endurance. The googolplex then, is a specific finite number, with so many zeros after the 1 that the number of zeros is a googol.

Incidentally, the name googolplex was originally proposed as the name of that famous internet search engine. It was shortened to googol and then misspelt as Google. The company with that name has only about 10^{15} bytes of data in its system.

A googolplex is 10^{googol} or $10^{10^{10^2}}$, which is a big number that is also difficult to write. Numbers like this are sometimes referred to as *power towers* because of the way the exponents stack upon each other. Since they are difficult to write, and because this really is just the tip of the iceberg for truly large numbers, there are several alternative notations. Here I will describe one: the *Knuth up-arrow notation*. It starts out looking simple:

$$10^{10^{10}} = 10 \uparrow\uparrow 3 \quad \text{or} \quad 3^{3^{3^3}} = 3 \uparrow\uparrow 4$$

Despite the simplicity of this notation these are already huge numbers. Since 10^{10} is 10 billion, $10 \uparrow\uparrow 3$ is a one followed by 10 billion zeros. Likewise $3 \uparrow\uparrow 4 = 3^{7,625,597,484,987}$, which is 7 trillion threes multiplied together, which yields a number with 3.6 trillion digits.

This double-arrow notation is clearly about pretty big numbers, and it also is a extension of exponents or powers. Also, it is clear where this notation came from. The single-arrow notations looks like:

$$10^{10} = 10 \uparrow 10 \quad \text{or} \quad 3^3 = 3 \uparrow 3 = 27$$

But what about a number like $3 \uparrow\uparrow\uparrow 2$? Double-arrows and triple-arrows are just arithmetic operations like multiplication and addition, just a bit more challenging. We might do well to step back and review what we think we know about arithmetic.

The first operation of arithmetic is addition:

$$3 = 1 + 2 \quad \text{or} \quad c = a + b.$$

The second operation of arithmetic is multiplication (subtraction is just negative addition):

$$6 = 2 \times 3 = 2 + 2 + 2 \quad \text{or} \quad c = a \times b = \underbrace{a + \ldots + a}_{b}$$

So we can write the second operation (multiplication) in terms of the first operation (addition) repeated many times.

The third operation of arithmetic is exponentiation:

$$32 = 2^5 = 2 \times 2 \times 2 \times 2 \times 2 \quad \text{or} \quad c = a^b = \underbrace{a \times \ldots \times a}_{b} = a \uparrow b$$

Again, exponentiation is the repeated application of the previous operation, namely multiplication.

The fourth operation is called the *tetration*, and brings us to our power tower:

$$3^{3^{3^3}} = 3 \uparrow\uparrow 4 = 3 \uparrow (3 \uparrow (3 \uparrow 3))$$

or

$$c = a \uparrow\uparrow b = \underbrace{a \uparrow (a \uparrow (\ldots a \uparrow a))}_{b}$$

At this point they are also usually called hyperoperations instead of just operations.

I will just mention the fifth hyperoperation to give us a feeling for how big these numbers are becoming. Consider:

$$3 \uparrow\uparrow\uparrow 3 = 3 \uparrow\uparrow (3 \uparrow\uparrow 3) = 3 \uparrow\uparrow (3^{3^3})$$

As we saw before, $3 \uparrow\uparrow 3$ has 3.6 trillion digits. That means that $3 \uparrow\uparrow\uparrow 3$ is a power tower with 3.6 trillion tiers going up!

The reason I have mentioned the up-arrow notation is that there are problems whose solution require it, much like the way Archimedes had to invent a number system to count those grains of sand. There is a famous problem in mathematics that involves cubes in higher dimensions and the lines that connect their vertices. Ronald Graham was working on this problem in the 1970s. He did not solve the problem, but he did put an upper bound on it. Martin Gardner, the mathematics writer for *Scientific American*, named this upper bound, *Graham's number*, and for many years it held the distinction of being, according to Gardner, the "largest number ever used in a serious mathematical proof."

Graham's solution started with the number $g_1 = 3 \uparrow\uparrow\uparrow\uparrow 3$, which is a hefty number already. But in the next step he blows that number away.

$$g_2 = 3 \underbrace{\uparrow\uparrow \ldots \uparrow\uparrow}_{g_1} 3$$

That is, a number with g_1 arrows. But it gets worse.

$$g_3 = 3 \underbrace{\uparrow\uparrow \ldots \uparrow\uparrow}_{g_2} 3$$

And the cycle continues until we get to Graham's number:

$$G = 3 \underbrace{\uparrow\uparrow \ldots \uparrow\uparrow}_{g_{63}} 3 = g_{64}$$

Despite its size, that number is finite. In principle, you can count that high and name that number.

***

One final word on combinations. In the study of dynamical systems there is something called the *Poincaré recurrence theorem*, which was alluded to when we looked at shuffled card decks. The theorem says that a dynamical system, given long enough, will return to a state very close to its initial conditions. This theorem has caught people's imagination because it seems to say that given enough time the universe will arrange itself, atom by atom, as it is now. And then history will repeat itself, exactly. Actually the theorem has a number of caveats that means it may not apply. Poincaré's theorem requires a constant *phase space*, which is a map of where things are and where they are going. In an expanding flat universe, this condition may not be met. Still, there was a recent calculation of the recurrence time for all the particles in the universe that arrived at:

$$\text{``}t_{\text{Poincaré}} \sim 10^{10^{10^{10^{2.08}}}} \quad \text{Planck times, millennia, or whatever."}$$

This is about $10^{10^{10^{100}}} = 10^{10^{\text{googol}}} = 10^{\text{googolplex}}$, a number with a mere googolplex digits. I like how the author wrote the time unit as "Planck times, millennia, or whatever." Planck time is 10^{-44} seconds, a millennium is a thousand years or 10^{10} seconds. In other words, being off by 54 digits does not really matter.

***

Yet there is something larger than Poincaré recurrence time for the universe or even Graham's number, and we use it every day. This is *infinity*. Since throughout this whole book I have plotted objects on a logarithmic scale, I will try to do that here too (see Figure 14.1). But even on a logarithmic scale, infinity is a long way off. If I take everything we have talked about and shrink them to a single pixel, infinity is off the page, and so the best I can do is draw an arrow to show that the line continues.

Right in the middle is a dot that is labeled "Plank length" (10^{-34} m) and "size of observed universe" (10^{27} m) and they appear on top of each other. That dot also contains numbers such as Poincaré recurrence time times the speed of light (to get meters) and Graham's number of meters (or Planck lengths, or parsecs). The axis stretches off to the right, off the

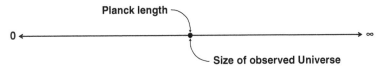

Figure 14.1 Everything is small compared to infinity. The Planck length and the size of the universe are on top of each other in a logarith[ ] tries to also show zero and infinity. Zero is still infinitely far [ ] infinity is infinitely far to the right.

page, beyond the Earth, beyond our galaxy and even beyond the edge of the observable universe. The other end of the axis also points off forever, even on a logarithmic scale, but not towards negative infinity. Rather, it points off to ever smaller and smaller sizes towards zero. Infinity and zero are somehow related.

One of the first times that infinity and the infinitesimal had a bearing on physics was in Zeno's paradoxes, which were mentioned in Chapter 8 when we talked about the flow of time. The most famous of Zeno's paradoxes was a race between Achilles and a tortoise. Zeno could have used a photon and continental drift to make his point, but instead he used Achilles, the athletic Greek warrior who chased Hector around the city of Troy three times in the Iliad.

In the paradox Achilles and the tortoise are to race. The swift Achilles gives the tortoise a head start, confident that he can easily catch him. A minute after the race starts Achilles has run to where the tortoise started, but the tortoise has moved forward a short distance. A few seconds later Achilles has moved to the tortoise's one-minute mark, but the tortoise has already moved forward a further finger's width. By the time Achilles has crossed the finger's width, the tortoise has traversed an additional hair's breadth, and so forth. Zeno was led to the conclusion that Achilles could never catch the tortoise; that it was logically impossible for him to do so. Therefore, Zeno summarized, motion was an illusion; a view supported by his mentor Parmenides.

Most physicists see this as a sort of calculus problem. An infinite number of infinitesimal steps leads to real, finite motion. In fact this is the basis of the *calculus* of Newton and Leibniz. We can treat a function or a trajectory as if we can divide it into an infinite number of parts, each infinitesimally wide. Whereas Newton's own opus, *Principia*, tended to not rely on calculus, it is hard to conceive of modern physics without it.

But what really is infinity? A good way of trying to get a handle on an abstract concept like infinity is to make a list of collections that have an infinite number of members (see Table 14.1). For instance, there are an infinite number of *natural numbers* (positive integers). There is also an infinite number of odd integers, an infinite number of even integers, an infinite number of perfect squares and so forth. It was Georg Cantor (1845–1918), a German mathematician, who first really figured out how to understand infinity.

First, a collection is called a *set* and the number of members of a set is called its *cardinality*. For example, the cardinality of the alphabet is 26. We can also talk about the cardinality of a set that has an infinite

Table 14.1 Examples of sets that have an infinite number of members, or cardinality of $\aleph_0$.

Set name	Members						Relationship
Natural numbers	1	2	3	4	5	...	x
Odd numbers	1	3	5	7	9	...	$2x - 1$
Even numbers	2	4	6	8	10	...	$2x$
Squares	1	4	9	16	25	...	x^2
Prime numbers	2	3	5	7	11	...	—

number of members. In fact the cardinality of the set of natural numbers is $\aleph_0$. This symbol is *aleph-naught*, aleph being the first letter of the Hebrew alphabet.

Cantor pointed out that in all these examples there was always a one-to-one relationship with the natural numbers. This means all of these sets have the same number of members, or cardinality, as the natural numbers. I do not actually need an equation, I just need to be able to order them. For example, according to Euclid's theorem there are an infinite number of prime numbers. Since I can match each natural number with a prime number, $1 \leftrightarrow 2, 2 \leftrightarrow 3, 3 \leftrightarrow 5, 4 \leftrightarrow 7,$ $5 \leftrightarrow 11, \ldots$ and so forth, there is a one-to-one relationship. The cardinality of all of these examples is $\aleph_0$.

What makes infinite sets unique is that one set, for example, the perfect squares, has the same cardinality as natural numbers, yet it is missing members. This does not happen with finite sets, and so it is a way of recognizing infinite sets.

Cantor went beyond this; natural numbers are just the starting point. The cardinality of *ordinal numbers* is $\aleph_1$, so these are a different type of infinite. The real numbers, which include *irrational numbers* such as π and $\sqrt{2}$ are continuous, and the cardinality of the *real numbers* is denoted as $2^{\aleph_0}$ or $\beth_1$ (beth-one where beth is the second letter of the Hebrew alphabet). It can also be shown that these are very different to, and much larger than, $\aleph_0$. Apparently there are an infinite number of types of infinity.

***

So how important is infinity to the universe? True, we use calculus to chart the smooth curve of a planet in orbit or the trajectory of an electron. But is that the way that nature really works? Remember, down at the Planck scale, at 10^{-34} m, we have reasons to think that nature might be broken into tiny, but finite bits. Real physical space and time might be quantized and not made up of the infinitesimal.

What is clear is that while the physical world may be bracketed by the Planck length and the edge of the observable universe, mathematics knows no such bounds. The mathematical universe is not only growing with each generation of curious minds, but it seems to me that the growth is accelerating into a truly limitless space.

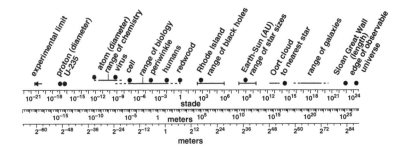

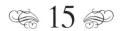

❧ 15 ☙

Forces That Sculpture Nature and Shape Destiny

I promised you at the beginning of this book that by the end we would be able to turn the question "How big is big?" into an answer: "This is how big 'big' is." The problem was not just learning a number like 4.3×10^{26} m, the distance to the edge of the universe, or 1.6×10^{-35} m, the Planck length, but really appreciating what that 10^{26} means. What is one hundred septillion?

We have tripped through analogies, skirted across various number systems, touched on similes and descriptions. But almost always we have come back to that logarithmic plot, the *scales of nature* plot. This is because I think we understand how big something is by comparing it to things that are a bit bigger or a bit smaller then it. We can say that there is a factor of 10^{42} between a proton and the size of the universe, but that is not how we picture them in our heads or how we understand these sizes. We need to take steps, going down in size—periwinkle, hair, cell, virus, molecule, atom and then that leap to nucleus—or going up—whale, redwood, kilometer, Wales, Earth, to the Moon, to the Sun. There are still big leaps, such as the one from an AU to the Oort cloud, or the jump from interstar distances to galactic scales. But the jump from the Sloan Great Wall to the size of the observable universe is only a factor of a hundred and not so hard.

We understand how big something is by where it is located on our scale, and what its neighbors are. So I think this one plot of scales becomes central to our understanding of "big." Therefore I would like to look at it one more time with a critical eye and ask if the way I have drawn it somehow distorts our view. Is there a bias? I will address this in two parts: first looking at the axes that are drawn and secondly looking at the objects that are plotted.

When drawing the axis, I have usually plotted object sizes expressed in meters and incremented major tick marks by a factor of ten. To see that this does not introduce a bias, at the start of this chapter I have plotted the same objects on three scales. The first scale is using the ancient unit of stades instead of meters. The second scale is the usual one, marked in meters. The final scale is also meters, but the tick marks are based on doubling, base 2, instead of factors of 10. This will change the words we use. The size of an atom is 5×10^{-13} stade, or 2^{-33} m. The numbers were simply multiplied by a conversion factor, but the spacing between objects did not change. No matter what scale we use, the solar system, as bounded by the orbit of Neptune, is still 6500 times bigger than the Sun. The relative spacing of the objects does not change and this comparison of objects is really what is important.

What about the choice of objects I have plotted? Have I omitted legitimate objects, or have I grouped together things that should not be seen as a class or a type?

Yes I have omitted real and interesting things. This book has not been encyclopedic and has not tried to be. This is especially true when dealing with the scales of our everyday life. I have not discussed the sizes of flowers and islands. I once read a proposal for a classification system that would label hills, mountains, massifs, ranges and so forth based on their heights, prominence and expanses. Humans have a nearly inexhaustible desire to organize things and to name them. That landscape naming system did not even start to address what are a fells, hummocks, alps, tors and so forth. But it is not clear to me that these are necessarily distinct and fundamental objects of the cosmos.

Finally, have I lumped things together that should not be in the same group? This is an ancient problem that even the Greeks wrestled with. I will, however, follow the lead of naturalist Carl Linnaeus, who wrote *Systema Naturae* (1735) and tried to organize all living things. He introduced a hierarchy to the taxonomy of plants and animals: kingdom, class, order, genus and species, based primarily upon the appearance of the reproductive part of a plant or animal. His system has been greatly modified over time, with the addition of phylums, families and subdivisions, but it was a starting point. The most important change came after Darwin. Classifications are now not based upon appearance, but on evolutionary history. We do not classify two objects that appear similar but have radically different histories as the same type. The Mona Lisa, and a perfect forgery, down to the last molecule, are not the same because they have different histories. One was painted by Leonardo da Vinci in about 1500 and the other yesterday. Among plants and animals there are a number of species in the new world and old world that have the same common name because, at least superficially, they look the same, for example the robin and the buffalo. When two species evolve a similar characteristic the process is referred to as *convergent evolution* or *homoplasy*; bats have wings, but they are not birds. Whales and dolphins are not fish, a mistake that Linnaeus made, but later corrected.

History and evolution can also be applied to inanimate objects, for example when recognizing what a planet is. In 2006 the International Astronomical Union (IAU) defined a planet as a celestial body that (a) orbits the Sun, (b) is effectively spherical and (c) has cleared its neighborhood. This last criterion is where Pluto fell short. This criterion is more about the history of Pluto than about the fact that it lives in a messy corner of the solar system. Pluto and its orbit were not formed in the normal way a planet forms.

How do we recognize a group of individuals as members of the same category? They should have similar appearance, but that is sometimes hard to quantify. They should be about the same size. They should have been put together in the same way; that is, they should have had a similar history. They should do similar things. In fact for much of the world we can recognize an object by what it is made of and what forces shape it. Galaxies are made of stars. Atoms are made of electrons, protons and neutrons. Planets and stars are both held together by gravity, but the difference is that in stars fusion is at play. In a star, the nuclear, strong and weak forces are at work transforming energy.

So I think that our diagram really does reflect nature and the variety of things in the universe. From quarks and protons to the great walls and voids of galaxies, all of nature is laid out here. Of course there are things missing, but the plot is already cluttered. Still, there is one significant gap. There is a void between 10^{-14} m and 10^{-10} m. There is nothing in nature in the interval between the largest nucleus (uranium-235, r = 7.75×10^{-15} m) and the smallest atom (helium, r = 0.31×10^{-10} m).

Here is a region of our diagram that is calling out for explanation. Why is it that there is a gap spanning four orders of magnitude?

***

Below 10^{-14} m all the objects of nature are held together by the strong force, above this gap none. The primary or fundamental constituents below the gap are the six types or flavors of quarks: up, down, charm, strange, top and bottom. These are bound together by the strong force and gluons. All of this—the characters and the rules of engagement—are described by quantum chromodynamics.

From this small collection of ingredients all of particle and nuclear physics arises. It tells us how nature builds protons, neutrons and pions, and occasionally more exotic particles such as the omega or charmonia (a combination of a charm and anti-charm antiquark). It also explains the nuclear force as a residual, a little bit of attraction that squeezes between quarks and tethers neighboring nucleons together. The nuclear force is what binds neutrons and protons to form a nucleus, the anchor of the atom.

On our scale of nature diagram, the region dominated by the strong force is well removed from the rest of nature. Only the strong force, living up to its name, is strong enough to bind particles into such a confined region. It also suggest that forces might be a useful organizing principle for the whole diagram.

***

The next object above the gap is the atom. Electrons and nuclei are the primary constituents of an atom and they are held together by the electromagnetic force. Much like the strong force in nucleons, the force between electrons and nuclei do not perfectly cancel each other out; they leave a residual, the chemical force. The electromagnetic force not only holds atoms together, it causes all of chemistry. Chemical bonds in turn form molecules, compounds, crystals and metals. It is chemistry that holds together macroscopic objects such as cells, plants and animals. The desk I am sitting at would collapse and those boulders in space we call asteroids would disintegrate if it were not for chemical bonds and the electromagnetic force. From atoms (10^{-10} m) to that irregular moon of Neptune, Proteus (2×10^5 m) the electromagnetic force and chemistry dominate.

As an aside, the name *electromagnetic* tells us a bit about how our understanding of this force has changed. At one time scientists viewed the attraction between charged particles and the force between magnets as two separate and unrelated phenomena. But James Clerk Maxwell (1831–1879) saw a symmetry between them and proposed that the two were actually flip-sides of the same coin, two facets of the same force. The magnetic force is caused by charges in motion. It might be electricity in a wire wound around an iron bar in an electromagnet, or an electron in orbit around an iron nucleus in a permanent magnet. This was the first force unification in the history of physics. In recent years one of the holy grails of physics has been to devise a *grand unified theory*, or GUT. This would be a theory in which the strong, weak and electromagnetic forces are three manifestations of a deeper underlying force. As of this time various GUTs have been proposed, but there is no consensus as to a final GUT.

Beyond GUTs there maybe a *theory of everything* or TOE, which includes gravity. But there is still a lot of work for physicists to do before either a GUT or a TOE is recognized as describing nature.

Still, we have made progress since Maxwell published his unification in 1865. In the 1970s Abdus Salam, Sheldon Glashow and Steven Weinberg showed that the electromagnetic and weak forces arose from the same set of fields, the electroweak fields. In the process they also predicted the existence of the W and Z bosons, which were experimentally confirmed a decade later. So when I use the term "electromagnetic" instead of "electroweak," I may sound a bit dated. But I am interested in persistent objects, things that are stable and have structure. We do not know of anything with a structure that depends upon the W or the Z boson.

***

The next and last force that sculpts nature is gravity. If we were to take all the moons and asteroids in the solar system and line them up by size, the smallest are simply rocks; even the moons of Mars look like potatoes. But as we look at larger and larger bodies there is a transition from irregular lumps of stone to well-formed spheres. Bodies as big as Miranda, a moon of Uranus (radius of 236 kilometers) or bigger, are spherical. Someplace at about 200 km radius we cross the hydrostatic equilibrium line and gravity is strong enough to pull down the high points and fill in the basins. Gravity has overcome the intrinsic strength of the rock.

To appreciate how, at one scale, chemical bonds can dominate yet at another scale gravity wins, imagine that you are standing on the Royal Gorge bridge, nearly 300 m above the Arkansas River in Colorado. Now take a steel jack chain and start to lower it over the side. Jack chain is actually a pretty weak chain. Each link is made of a piece of wire that is twisted like a figure eight. It keeps its shape because of internal chemical forces. Our chain is #16 gauge, made from wire about 0.16 cm in diameter. So after a 100 m of it have been reeled out there is almost 7 kg of chain hanging from the bridge. The breaking point of this chain is about 18 kg and so we can keep reeling it out. After 200 m, 14 kg hangs below us. Someplace near 250 m we can expect one of the links to straighten and the chain to break. In fact, if the links are identical, and we reel it out smoothly, we would expect the top link to break since it bears the most weight. Gravity has overcome chemistry and most of our chain plunges into the river below. (Most chains would be a few kilometers long before breaking, but bridges are not that high.)

Gravity is the weakest force in nature, yet in the end it is the force that shapes the largest objects: planets, stars, galaxies and even the large-scale structure of the universe. This is because of one unique trait of this force: gravity has only one charge; there is only mass and no anti-mass. Gravity always attracts. Two quarks with the same color charge repel, two electrons with the same electric charge repel, but particles with mass always attract.

If I had 1 g of hydrogen and I could separate the electrons and the protons into two containers a meter apart there would be a force equal to about a billion tons trying to get them back together. Normally we do not feel that force because the positive protons are paired with the negative electrons on a microscopic level, canceling each other. That sample of hydrogen also has quarks with color charge and tremendous forces between them, but again they cancel on human scales and we are not aware of them. That sample also has mass, a single gram, but the force of that gram, although extremely weak, is felt across the universe.

***

On our scales of nature diagram, we can now group things by the force that primarily shapes them (see Figure 15.1). Below the 10^{-12}-m gap

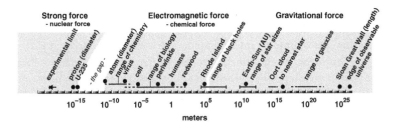

Figure 15.1 Different forces of nature dominate different scales. The *scales of nature* diagram, with objects plotted by size from our experimental limit of smallest object, to the edge of the universe. Regions where different fundamental forces dominate are indicated.

the world is dominated by the strong force; from the atom to Proteus is the realm of electromagnetic and chemistry; above that is gravity. Is this significant? Does it point to some deep underlying principle? I do not think so. There are too many exceptions and too many situations where more than one force is important. Why is the tallest tree on Earth, the Hyperion, limited to 115 m? Gravity has limited trees' height, even though trees are well below the size of spherical moons. Also, black holes exist well below the size of irregular celestial bodies. The limit of the largest stable nucleus, uranium-238, is set because the neutrons and nuclear force can no longer overcome the electric-charge repulsion. In fact the limits of the sizes of different objects is often set by a balance between forces, but combined in a unique way for that situation. This diagram may not reveal anything deep, but it does help us organize what we see.

***

There remains one glaring problem with this size-to-force relationship: we have omitted one of the fundamental forces of nature, the weak force. The weak force is associated with radioactive decay. In its simplest form a neutron decays into a proton as well as an electron and an antineutrino. When cobalt or radium undergo a radioactive decay this is exactly what is happening inside the nucleus. In fact, if we were to peer deep inside the neutron we would see that a down quark is decaying into an up quark. These are rare events. The lifetime of a neutron, if it is not bound in a nucleus, is between 14 and 15 min. For comparison, the delta particle decays via strong interaction with a lifetime of 5.58×10^{-24} s.

Perhaps more important in shaping the world as we know it is the reverse but even more difficult process: when a proton decays into a neutron, positron (anti-electron) and neutrino. That reaction does not take place spontaneously like a neutron decay. Because a neutron is heavier than a proton you need to put energy into the system. However, it can happen when two fast-moving protons collide: one proton becomes a neutron, using some kinetic energy and the neutron and the other proton fuse to become a deuteron (see Figure 15.2). This releases more energy than it used, and causes the stars to glow. At the heart of the process that powers stars and lights up the universe, up quarks are changing into down quarks and that is the weak force at work.

After that initial fusing things move faster. The positron and any nearby electrons will collide, annihilate and give up even more energy. The deuteron will combine with hydrogen to form helium-3 and yield even more energy. But what sets the pace for everything is that first step. A proton can bounce around inside a star for a billion years before that occasional weak decay happens at just the right moment.

The weak force has little to do with the shape and spatial structure of any object in nature. But it has a huge amount to do with the pace at which it evolves. The weak force governs how fast stars burn and

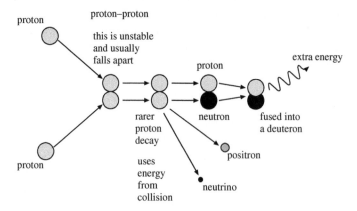

Figure 15.2 Proton–proton fusion. This process can only happen if one proton decays into a neutron and that transformation is due to the weak force. The weak force sets the rate at which stars burn hydrogen.

stretches their lifetimes out to billions of years. If the combining of two protons to form a deuteron took place via the strong interaction it would happen a quadrillion times faster. It would be reasonable to expect stars to burn hotter and faster. If this were true the universe might still be expanding but it would look a very different place. All the fuel would be used up by now and the stars would be dark.

Since the weak force sets the pace of energy use, our Sun has burned for billions of years and will continue for a few billion more. The weak decay also sets the brightness of our Sun and so the temperature on the surface of our planet. And that, at least from the point of view of humans and life on Earth, is pretty important.

***

So again, do we understand the size of the universe or the size of quarks? Can we comprehend the lifetime of a delta particle or the age of the universe? Scientists who spend their lives studying one of these systems can see with their eyes no more than anyone else: a hairsbreadth at the smallest scale and the Andromeda galaxy, if they know where to look, at the largest. But if they truly understand the scale of their subject it means that they know how to manipulate the numbers that are relevant and ignore the ones that are not. The marine biologist who is studying Baleen whales does not worry about the effects of black holes, the physicist looking for glueballs is not concerned with carbon bonds, and the astronomer measuring the distance to the Sloan Great Wall may not be concerned about any of these. When we say that a scientist understands how big or how small something is, it means they know how it relates to other things of similar magnitude that have influence upon their subject, and that they can apply that knowledge successfully.

We stand on a planet 10^{22} fm or 10^{41} Planck lengths in diameter, made out of 10^{52} quarks and we are giants. We live in a universe 10^{27} m, or human lengths, across and we are insignificant iotas, except for one thing. We can look at stars and cells and atoms and galaxies and, even if our understanding is incomplete, we can comprehend them all. We can take in all the scales of nature and marvel at the world.

Index

A

accelerator, 169
Adam, Jehan, 31
Almagest, 54
analemma, 114
Andromeda galaxy, 16, 218, 219
anemometer, 46
angstrom, 7, 154
Ångström, Anders, 154
antimatter, 175
antiquarks, 175
Arabic numerals, 81
Archimedes, 80, 209
Avogadro's number, 91
number system, 91
sand in universe, 229
argon dating, 141
Aristarchus of Somos, 86
asteroid belt, 201
astrolabe, 113
astronomical unit (AU), 7, 189,
194
atom, 242
illustrated wavefuctions, 161
size of, 155
spectrum, 122
atomic mass unit (u), 41
Avogadro's number, 36, 43
Avogadro, Amedeo, 36

B

bacteria, 18
Balaenoptera musculus, 26
baseball, 118
baseline, 87
Beaufort scale, 46
Beaufort, Francis, 46
Bernoulli, Daniel, 68

Bessel, Friedrich, 210
Big Bang, 117, 144, 145, 223,
226
billion, 30
binomial system, 7
black holes, 214
blackbody radiation, 72
Boötes void, 225
Bohr, Niels, 159
Boltzmann's constant, 71, 73
Boltzmann, Ludwig Eduard, 68,
230
Bradley, James, 210
Brahe, Tycho, 209
Breguet, Abraham-Louis, 130
bristlecone pine, 14
BTU, 102

C

Calculus, the, 236
caloric, 64
calorie, 102
Cambrian revolution, 141
candle, 220
canonical hours, 128
Cantor, Georg, 236
Capricornus void, 225
carbon-14 dating, 136
cardinality, 236
cards, 230
Carnot cycle, 65
Carnot, Nicolas Léonard
Sadi, 65
Carroll, Lewis, 78, 80, 227
Cassini, Giovanni Domenico,
188, 189
cells, 16, 21
Cepheid variables, 218

CERN, 179
cesium-fountain clock, 117
Chapman, Aroldis, 118
chemical
    vs. nuclear, 106
    bonds, 154, 196
    vs. gravity, 244
chemistry, 153–155
Chinese, traditional time, 127
Chuquet, Nicolas, 31
Clausius, Rudolf Julius Emanuel,
    66
clocks, 112, 116
clonal colonies, 28
comet, 133
continental drift, 142
Copernicus, Nicolaus, 189, 190
cosmic distances
    astronomical unit, 189
    candle, 220
    Cepheid variables, 218
    Hubble's law, 224
    spectral parallax, 213
    stellar parallax, 209
    Tully–Fisher, 220–221
    type Ia supernova, 221
cosmic ladder, 213
Crab nebula, 214–215
crystals, 156
cubit, 3

D
Dalton, John, 36
day, 112, 131
decibels, 58
decimal time, 128
Delambre, Jean Baptiste Joseph,
    5, 147
delta particle, 123, 126
delta time, 116
diamond, 156
diatonic, 60

dinosaur
    extinction, 141
    size of, 26
DNA, 17
Doppler effect, 220, 221
dwarf planet, 201–203

E
Earth
    age of, 143
    size of, 84
    spherical, 83
earthquake, 50
Eddington luminosity, 207
Einstein, Albert, 75, 223
electromagnetic force, 108, 135,
    242
electroweak, 243
elephant, 24
energy, 97
    chemical, 104
    conservation of, 66, 97
    food, 101
    fuel, 102
    gravitational, 102
    nuclear, 102
    tidal, 100
    wood, 102
    meteor, 103
    nuclear, 106
    potential, 98
entropy, 66, 69
equation of time, 114
Eratosthenes of Cyrene, 3, 84
eukaryotes, 18
eyes, 14

F
Fechner's law, 57
Fechner, Gustav, 57
fen, 127
fermi, 7, 170

Fermi, Enrico, 170
Feynman diagram, 177
Feynman, Richard, 177
foil, 149
football (soccer), 33, 95, 118
force
    electromagnetic, 108, 135, 242
    four fundamental, 107
    general features, 171
    gravitational, 108, 135, 197, 243
    nuclear, 165, 170, 171, 177
    strong, 108, 135, 174, 242
    weak, 109, 135, 245
frames per second, 118
François, André Méchain Pierre,
    5, 147
Franklin, Benjamin, 153
French Republic calendar, 128
French revolution, 128
fusion, 144, 246

G
GAIA space telescope, 212
galactic year, 134
galaxy, 205
    barred spiral, 216
Galilean moons, 184–186
Galilei, Galileo, 24, 184–187
Gardner, Martin, 234
Gay-Lussac, Joseph Louis, 37
Gelon, King, 80
General Conference on Weights
    and Measurement, 147
general relativity, 120, 223
geologic time, 138, 144
geothermal, 100
GeV, 169
gigahertz, 120
globular clusters, 217
    age of, 144
gluons, 172
gnomon, 112

Google, 232
googol, 231
googolplex, 232
Graham's number, 234
Graham, Ronald, 234
Grand Canyon, 87, 139
grand unified theory, 243
gravitational
    force, 108, 135, 196, 243
    lensing, 214
    vs. chemical, 244
Great Nebula of Cerina, 215
Great Wall, 225
Greek number system, 81
guitar, 62
Gutenberg, Beno, 49

H
hair, 15
    measure, 150
Hale-Bopp, comet, 133
Halley's comet, 133
Halley, Edmond, 133, 190, 210
Hamlet, 231
harmonics, 61, 160
Haussmann, Baron, 21
heat, 64
Heisenberg uncertainty principle,
    121, 171
heliometer, 55
heliosphere/heliopause, 202
helium-neon (HeNe) laser, 121
Henderson, Thomas, 210
heptatonia prima, 60
Hertzsprung, Ejnar, 212
Hertzsprung–Russell diagram,
    212, 213
Hindu–Arabic numeral system,
    81
Hipparchus of Nicaea, 54
Hipparcos space telescope, 212
Hirst, David, 118

Hofstadter, Robert, 170
homeothermic, 24
horology, 116
horsepower, 96
Hubble's law, 223
Hubble, Edwin Powell, 218, 223
Huygens, Christiaan, 208
Hyakutake, comet, 133
hydrogen, 122
hydrogen burning limit, 207

I
infinity, 235–237
interference, 150, 152, 157
island universe, 16, 218
isotope, 165

J
joule, 96
Joule, James Prescott, 97
Jovian planets, 199

K
Kant, Immanuel, 16, 218
Kasner, Edward, 231
ke, 127
Kelvin, 33
Kepler's laws, 132, 134, 190, 191
Kepler, Johannes, 132, 185
kinetic theory of gasses, 67
Kirkwood gaps, 201
Knuth up-arrow notation,
    232–234
Kuiper belt, 202

L
Large Hadron Collider, 179
large-scale structure of the
    universe, 224, 225
law of definite proportions, 36
laws of thermodynamics, 66
Leavitt, Henrietta Swan, 218

lifeimes, 135
light, *see* speed of light
    aberration, 210
    wavelength measure, 150
lightning, 119
lightyear (ly), 7, 31, 211
Linnaeus, Carolus, 7
Littorina, 22
local group, 219
local magnitude scale, *see* Richter
    scale
logarithm, 48
logical positivist, 72
long scale, 31
longitude, problem, 187
Loschmidt's paradox, 69
Loschmidt, Johann Josef, 39, 68
Loxodonta, 25
luminosity, 56

M
Mach, Ernst, 71
magnitude
    absolute, 212
    apparent, 212
Mathematics and the Imagination,
    231
Maxwell, James Clark, 39, 67,
    243
Maxwell–Boltzman distribution,
    68
Mercalli intensity scale, 50
Mercalli, Giuseppe, 50
meter, 4, 6, 146, 148, 240
    Earth based, 147
    krypton-86, 148
    light, 148
    platinum–iridium bar, 147
metre, *see* meter
miao, 127
microstates, 71
mile, 3

Milky Way, 134, 216–217
    name, 205
Mohs's scale of mineral hardness,
    52
Mohs, Friedrich, 52
molecule, 155
    size of, 39, 153
molecules-in-ocean problem, 34,
    43
monkeys typing, 231
Moon, 131
    distance to, 86
moons, 195–197
moons of Jupiter, *see* Galilean
    moons
Munich arsenal, 65
muon, 119
musical scales, 60

N
names of numbers, 31
nanosecond, 120
natural units, 76
nebula, 214–216
nerve pulses, 118
neutron, 165, 178
    lifetime, 131
    size of, 170
Newman, James R., 231
Newton's law of gravity, 134
Newton's laws of motion, 67
niche, 29
nuclear
    chain reaction, 120
    energy, 100, 106
    force, 165, 170–172, 177
    vs. chemical, 106
nucleon, 165
    size of, 170
nucleosynthesis, 144
nucleus, 165
    size of, 158

O
oil, 153
omega particle, 123
Oort cloud, 202
orbital resonance, 200
orbits, 158
Ordovician period, 110
Orion nebula, 215
overtones, 61, 160

P
Pangaea, 142
paper thickness, 149
parallax, 90
Parmenides of Elea, 112
parsec (pc), 7, 210
particles, 123
pendulum clocks, 113
pennies, 70, 230
periodic table of the elements,
    162
periwinkle, 22
phanerozoic eon, 140
photometry, 55
photons in the universe, 230
photosynthesis, 99
piano, 60
pion, 123, 170, 177
Planck length, 76, 181, 235
Planck's constant, 73, 76, 121
Planck's law, 73
Planck's relation, 75
Planck, Max Karl Ernst Ludwig,
    72
planet, 132, 198–201
    definition of, 201, 241
planetary rings, 197, 198
Pluto, 201
Pogson ratio P_0, 56
Pogson, Norman Robert, 56
Poincaré recurrence theorem,
    234

pool, 69
positional number system, 81
potassium–argon dating, 141
power, 96
power towers, 232
Preseli hills, 110
prokaryotes, 18
protein, 17
proton, 165
    size of, 170
psychophysics, 57
Ptolemy, Claudius, 54
public land survey system, 9
pyramids, 33
Pythagoras, 83

Q
quantum chromodynamics, 174,
    242
quantum mechanics, 74, 121, 159
    wavefunction, 160, 169
quarks, 172, 177, 178
    flavors, 175
    in the universe, 229

R
Rømer, Olaf, 188
radioactive decays, 136
Rayleigh, 153
redwood, 27
revolutionary calendar, 128
Richer, Jean, 189
Richter scale, 49
Richter, Charles, 49
Robinson, John Thomas
            Romney, 46
Roche limit, 197, 198
Roche, Édouard, 197
Roman number system, 81
rotifers, 21
Rumford, Count, 64
Russell, Henry Norris, 212

S
Saffir, Herbert, 47
Saffir–Simpson hurricane scale,
    47
Salisbury Plain, 110
sand in universe, 91
Sand Reckoner, 80
satellite, 186
scales of nature, 239, 244, 245
Schrödinger, Erwin, 17, 33
scientific notation, 35
sclerometer, 52
seeing, 168
seismograph, 49
sequoia, 27
shake, shakes of a lamb's tail, 120
shichen, 127
Shoemaker-Levy 9, comet, 133
short scale, 32
sidereal time, 114
Sidereus Nuncius, 185
Simpson, Bob, 47
Sloan Digital Sky Survey, 225
Sloan Great Wall, 225
Smith, William "Strata", 140
soccer, 33, 95, 118
solar
    cells, 99
    radiation, 99
    time, 113
    vs. sidereal time, 132
solar system, 198–204
    age of, 143
sound, 58
special relativity, 119
spectral parallax, 213
spectrum lines, 122
speed of light, 186
    Rømer, Jupiter, 188
    Galileo's measurement, 186
spherical vs. irregular bodies,
    195–197

sponges, 21
stade, 3, 85, 240
standard model, 180
star(s), 206–208
    age of, 144
    biggest, 207
    distance to, 208
    in the universe, 229
    maximum size, 207
    minimum size, 207
    nearby, 211
Starry Messenger, 185
statistical mechanics, 68
Stefan, Joseph, 68, 73
Stefan–Boltzmann law, 73
stellar
    magnitude, 54
    parallax, 209–210
Steno, Nicholas, 138
stomata, 22
Stonehenge, 110, 133
stratigraphy, 139
string theory, 182
strong force, 108, 135, 174, 242
Struve, Friedrich Georg Wilhelm
        von, 210
Sun, 206
    distance to, 87
    energy, 98
    magnitude, 58
    size of, 207
sundial, 112
supercontinents, 142
supernova, 144, 215, 221
surface to volume, 19
survey, 87
synodic period, 132
Syracuse, 80

T
table of nuclides, 166, 167
terrestrial planets, 199

theory of everything, 243
thin film, 153
Thompson, Benjamin, see
        Rumford, Count
time
    nanosecond, 120
    tidal effect, 116
time's arrow, 112
Titus-Bode law, 200
towns vs. city, 20
township and range, 8
trachea, 22
transit of Venus, 190–194
Tresca, 149
Tresca, Henri, 148
Triangulum galaxy, 16, 58, 219
Tully–Fisher, 220–221
Two New Sciences, 24, 186

U
universe
    age of, 144, 145
    photons in the, 230
    quarks in the, 229
    sand in, 91
    size of, 226, 235
    size of, Archimedes, 89
    stars in the, 229
University of Vienna, 68
uranium-238, 164

V
vanadium-50, 136
Virgo supercluster, 222
Voyager, 202

W
Walrus and the Carpenter,
        78, 79
warm blooded, see
        homeothermic
watt, 96

Watt, James, 96
wavefunction, *see also* quantum
    mechanics
wavelength, 150
waves, 160
weak decay, 135
weak force, 109, 135, 245
Weber–Fechner's law, 57
Wegener, Alfred, 142
whale, 26
Wien, Wilhelm, 73

wind speed, 46
WMAP cold spot, 225

Y
yoctosecond, 126
Yukawa, Hideki, 171

Z
Zeno's paradoxes, 236
zero, 235
Zeuxippus, 80